A Biodynamic Manual

A Biodynamic Manual

Practical Instructions for Farmers and Gardeners

PIERRE MASSON

Floris
Books

Revised and edited by Pierre and Vincent Masson

Translated by Monique Blais

Additional editorial work by Bernard Jarman

First published in French as *Guide Pratique Pour L'Agriculture Biodynamique*
by Éditions BioDynamie Services, Château in 2012
First published in English by Floris Books, Edinburgh in 2011
This edition 2014. Third printing 2021
© 2012 Editions BioDynamie Services
This translation © 2011, 2014 Floris Books

MIX
Paper from
responsible sources
FSC® C007785

British Library CIP Data available
ISBN 978-178250-080-3
Printed and bound in Great Britain by Bell & Bain, Ltd

Floris Books supports sustainable forest management
by printing this book on materials made from wood that
comes from responsible sources and reclaimed material

Contents

Geography and measurements

This book was written for France, for northern hemisphere climates ranging from temperate to Mediterranean. Obviously in different parts of the world things have to be adapted. In particular, readers in the southern hemisphere will need to adapt the months and seasons.

All measurements are given in metric and US measures, not imperial. Imperial measures are practically the same as US measures, except for capacity where the US gallon is 3.8 litres while the imperial gallon is 4.5 litres.

Photographic acknowledgments

Photographs by Pierre Masson, Vincent Masson, Jean Marie Defrance, Bruno Follador, Domaine Guillemot-Michel, Fréderic Lafarge, Adeline Tellier, Hélène Thibon, Adriano Zago.

Preface

This book is a guide to the application and fundamental practices of biodynamic methods. It is meant for professional farmers and gardeners alike who wish to start on, or advance along, this agricultural path.

The eight lectures known as the 'Agriculture Course' given in 1924 by Rudolf Steiner are the foundation of biodynamic agriculture. Renewing our approach to nature through agricultural activity and nutrition, these lectures provide the philosophical basis of the methods as well as many practical elements. The lectures are a valuable resource both for those wanting to use biodynamics in practice, and those seeking to deepen their knowledge of biodynamic farming and gardening.

From this starting point, a number of successors have written books extending the ground work. Among them are Ehrenfried Pfeiffer, Harald Kabisch, Maria Thun, Léo Selinger, Friedrich Sattler, Manfred Klett, Eckard von Wistinghausen, Volkmar Lust, and many others.

This book was inspired by a few newsletters for the Bio-Dynamic Agricultural Association of Australia by Alex Podolinsky. These pages not only reflect Podolinsky's practical methods, but are also the fruit of over forty years experience as a farmer and as a consultant for biodynamics in its diversity of husbandry, multi-crop grain production, viticulture, arboriculture and gardening.

It does not, of course, cover the entire spectrum of biodynamic methods, nor its fundamentals. Describing the methods of making the different preparations, as well as understanding the processes, takes time. The same goes for various practices commonly used in biodynamic practice, such as dynamic crop rotation, companion planting, and subtlely balancing a farm's environment. However, what we can find in this book are the essentials to get started, or deepen, our practice.

PIERRE MASSON

Preface to the Third French Edition

This third edition has been extensively updated by Pierre and Vincent Masson.

The book has been reorganised into three parts. Part 1 covers the practical essentials of biodynamic agriculture, such as biodynamic preparations, composting, working with manure, and working with rhythms.

Part 2 addresses further practical aspects such as using tree pastes, plant extracts, herbal teas and ferments, and pest and weed control. You'll also find guidance on products for stimluating and regulating plant health in this section.

Part 3 looks at specialist issues: green manure, seeds, animal husbandry, large-scale crop farming, vegetable farming, arboriculture and viticulture.

Thanks goes to many individuals for their help with this new edition: Jean Luc Petit, Frédéric Cochet, Jean-Louis Keller, Michel Leclaire, Alain Regnault, Roger Raffin, Philippe Fourmet, Benoit Massé, Reinout Nauta, André Ollagnon, Dr Bruno Giboudeau, Jacques Fourès, François Duvivier, Marc Guillemot and Frédéric Lafarge. Thank you also to the many farmers and growers who have shared their experiences and knowledge with us, helping to make this a much better book.

PIERRE AND VINCENT MASSON
EASTER 2012

11

Introduction

There are several principles which characterise biodynamic agriculture. Farms and other agricultural organisms should be fully integrated in both their physical environment (terroir) and their cosmic environment, which leads to health, balance and sustainability. Things which can help with this include:

- Sustainability of resources, including soil, plants and animals
- Respecting the natural habits of animals
- Application of horn manure (500 or 500P) and horn silica (501) and use of the six biodynamic preparations
- Following the rhythms of nature and of the cosmos
- Aiming for a sustainable level of productivity which does not affect the balance and health of the land
- Aiming for as great a diversity of plants and animals as possible
- Formation of living soil

In order for agriculture to be sustainable, the soil must attentively cared for. Not only does the soil need to be respected and conserved, it must be enlivened, regenerated and developed. This is an ongoing endeavour. Through the miracle of photosynthesis, the activity of the sun, and the formative forces of the universe, plants create living substance from the mineral kingdom. Drawing carbon dioxide and nitrogen from the atmosphere as well as from water, the plant enlivens the soil by its root secretions (rhizospheres), and in living soil this activity produces an increase in humic levels (microbial humus).

It is important to emphasise that biodynamic agriculture can only create a new relation between the earth and the cosmos, through the plant and through good agricultural practice. Furthermore, the application of biodynamic preparations cannot substitute or be separated from such good practices as:

- ❧ Proper management of organic matter and compost
- ❧ Working the soil when conditions are right, using appropriate tools
- ❧ Long term and diversified crop rotation
- ❧ Using seeds and plants adapted for local conditions
- ❧ Ensuring that the soil is never bare
- ❧ Proper management of cover crops and pasturing

The human being is at the heart of biodynamic agriculture. The farmer's quality of observation, the ability to assess a situation and the ability to act, make a farm successful and keep it sustainable.

Above all there needs to be a personal commitment to the work done on the farm: someone needs to take responsiblity for it. There also needs to be a continual seeking for a deeper knowledge of life processes, as well as an empathy with the living world around us and the forces that act on it. Biodynamic agriculture is a science of adaptation, of customising methods and processes.

We must never forget that the fundamental aim of biodynamic agriculture is the production of quality nutrition, whilst respecting nature's kingdoms and the creatures that live in it.

The earth, nutrition and human beings are therefore completely intertwined.

There is also a social dimension to biodynamic agriculture that should not be overlooked. It is inappropriate to work out of a sense of competition or materialism. The ideal, rather, is for a kind of economic community to be developed. This could begin with active participation in local meetings and work groups, addressing both production and consumer issues.

To get the most from this manual, it is recommended to participate in a seminar or workshop on biodynamic farming, and have the opportunity to take part in the stirring of preparations, and making a compost pile with an experienced biodynamic practitioner. Contact your local or national biodynamic association (see *Biodynamic Associations* at the end of the book).

Biodynamic produce is often sold under the brand name of 'Demeter'. Only strictly controlled and contractually bound partners are permitted to use the brand name. A comprehensive verification process ensures strict compliance with the International Demeter Production and Processing Standards, as well as applicable organic regulations in the various countries.

～ PART 1 ～

Core Biodynamic Practice

Mature horn manure after coming out of the horn

1. Horn Manure (500 and 500P)

Horn manure preparation, also known as the 500, is obtained from good quality cow manure. The manure is put into cow horns and buried in the soil over the winter period, allowing a transformation of the preparation underground. There are particular procedures and it's best to learn how to properly fill the horns before attempting it yourself. It is also possible to obtain good quality preparations from local biodynamic associations or to participate in their preparation-making days; or simply to approach neighbouring biodynamic farms.

The important thing is to use a preparation that has been completely transformed in the ground over the winter period. It should be moist, colloidal in nature (very fine, supple and elastic), have a brownish-black colour, and be without odour or with a very faint humic smell. Preparations that have started to dry out or have gone mouldy or rotten do not give good results.

This preparation, when it has been well prepared, stored and applied, works primarily on the soil and roots of plants.

- ⚘ It strongly helps build soil structure.
- ⚘ It stimulates microbial activity and the formation of humus thus improving the absorption and retention of water in the soil.
- ⚘ It regulates the acidity of the soil, increasing the pH for acidic soil and lowering it for alkaline soil.
- ⚘ It stimulates growth of the root system, particularly promoting growth to greater depths.
- ⚘ It increases the germination rate of seeds, the development of legumes and fabaceae and the formation of their nodules.
- ⚘ It helps dissolve hard pans, even at depth, and can counter salinisation.

Prepared horn manure (500P)

First tried and tested by Alex Podolinsky in Australia, prepared horn manure (500P) is horn manure (500) in which all six preparations normally meant for compost have been added (see chapter *Biodynamic Compost*). The synergy created makes 500P significantly more effective than the traditional 500 preparation. However, 500P should only be made by experienced biodynamic practitioners who have been specifically trained, so detailed instructions are not provided here.

Prepared horn manure (500P) has been proven to be effective on field crops, pasture, large market gardens, orchards and vineyards. It is suitable for use everywhere and replaces

Soil comparison: organic soil (left); soil treated biodynamically for a year (right). Observe the different in colour, soil structure and root development

Soil comparison: organic soil (left); soil treated biodynamically for five years (right)

500 in most cases. One or two applications per year are, in general, enough: once in spring and another in autumn. Two applications a year benefit soil and root development and are less time-consuming than the traditional method of three passes of autumn horn manure followed by springtime sprays of 500.

It should be stirred for one hour, and sprayed in the same way as the traditional horn manure preparation. It requires strict storage conditions and the quality and temperature of water used, as well as the shorter delay after stirring, are critical.

Laying out the cow horns, the openings always facing down to avoid trapping rainwater

Storage

Horn manure (500) is a living substance, and to maintain its efficacy, it needs to be carefully stored in a container inside a box as described below. Even purchased preparations must be handled properly as soon as they arrive: remove them from the packaging and store them as described. If you're not able to make a suitable box, you can often purchase them from your local biodynamic association.

Two sizes of storage boxes are described; the size of the box depends on the area of land. The small box is for areas up to 2.5 ha (6 acres) where less than 250 g (½ lb) of horn manure is needed. The large box holds 3 to 6 kg (6–14 lb). The interior space should be proportional to the size of the container: don't put a small jar in a large box.

Use a glass, stoneware or ceramic jar. Enamel can be used, but ensure there are no chips or flaking. Do not use clay pots fired at low temperature since these are often coated with a metal-based glaze which is not good for living substances. Do not use porous flowerpots that allow evaporation as this diminishes the activity of the preparations.

For **smaller quantities,** take a small glass jar, a jam jar, or similar. A lid is useful for small quantities that would otherwise dry up quickly. Ensure there is no contact with metal parts or synthetic seals. A plate or glass lid, without metallic rings, is best. The seal on the jar should never be completely airtight; the preparations need to breathe lightly.

For **larger quantities** (3 to 40 litres/quarts) use an earthenware pot, enamel or glazed crock with straight, upright sides. There must be one, or if possible several, drainage holes 5–8 mm (¼ in) diameter along the bottom. The holes should be covered with bits of cow horns that have been cut lengthwise. The container should sit on a non-porous saucer or plate to prevent the bottom of the storage box from becoming damp, which would cause the peat to lose its insulating potential. A plate or a plank of wood or slate can also be used to cover the container, but make sure that there is sufficient ventilation.

In **very large containers**, a layer of empty cow horns can be spread on the bottom. It is necessary to have a hole in the centre large enough to allow for effective drainage. This drain must be porous enough to allow the preparations to breathe.

The container storing the preparation must be placed in a double-walled box, surrounded by dry peat on all sides, including the double-walled top (ideally 6 to 8 cm, about 3 in). The peat must be adequately dried before use, since humid peat does not insulate well; additionally, it can rot the box and cause the wood to become swollen. Spread it out in thin layers and dry it in the sun or in a well ventilated room for several days or weeks, depending on its initial humidity and the season.

Use natural sphagnum peat, without any additional fertilisers, preferably originating from a country that has not accumulated too much radioactivity (ask for a certificate of analysis indicating the absence of radioactivity).

The peat within the storage case and lid should be checked occasionally (every 3 to 4 years). Since it has a tendency to compact, it can leave gaps that result in poor insulation.

Preparations must never be in direct contact with the peat; even peat dust can be detrimental. Be especially careful when handling the preparations.

The case should be stored in a cool, well ventilated and quiet area. A basement without odours is a good location. A roofed place protected from freezing is also suitable.

Avoid fuel smells, car exhausts, electrical outlets or electromagnetic fields from computers, motors, etc. Odours from essential oils (terpenes), sulphuric emissions and fermentation gases are harmful. Avoid fluorescent lights, as well as vibrations from a busy road, train tracks, machines, etc, and be especially aware of electro-magnetic pollution such as wi-fi and mobile (cell) phone signals.

When in storage, the preparation should be checked regularly, particularly in the springtime, after it comes out of the ground. (It is important to form a personal interest in all aspects of compost, as Steiner indicated in lecture 4 of the Agriculture Course.) Initially check on the preparation several times a week, then weekly and eventually check about twice a month.

If the preparation is too moist, aerate it as quickly as

possible by spreading it out on screens or a cotton sheet for several hours; this must be done in the shade. If it is too dry, several drops of good quality water (rainwater or a soft spring water) can be added to moisten it. Check after several days and repeat if necessary.

The consistency of the preparation should be the same as when it was dug out of the ground: colloidal humus (fine, supple and elastic) comparable to fresh earthworm droppings. Letting the preparation dry out greatly reduces its efficacy. But it must not be too damp as this can lead to anaerobic conditions and putrefaction, especially at the bottom of the container.

The presence of red worms (compost worms) is normal. If they become too abundant, you can take them out and return them to nature or to a compost pile.

Small storage box

This box is suitable for quantities of horn manure corresponding to areas between 0.1 and 2.5 ha (¼ to 6 acres) (up to 250g (½ lb) of horn manure).

Cover: 28 x 28 x 9 cm (11 x 11 x 3½ in). Fill the space with a layer of well-dried peat.

Box: the inner compartment, 10 cm square and 12 cm deep (4 in square, 5 in deep) can be made up of slats, much like hardwood flooring. The dimensions of the exterior should be 28 x 28 cm, and 21 cm deep (11 x 11 x 8 in). At least 6 cm (2½ in) of peat is necessary all around to line the inner compartment and the lid.

Leave a protruding screw in the top of the box to ensure a gap of 1 to 2 mm (¹⁄₁₆ in, or matchstick thickness) between the box and the cover to allow ventilation.

Storage box for horn manure

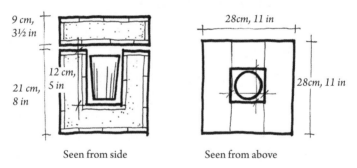

Seen from side   Seen from above

Large storage box

This box for is suitable for 3 to 6 kg (7 to 14 lb) of horn manure.

Cover: 46 x 46 x 9 cm (18 x 18 x 3½ in). Fill the space with a layer of well-dried peat.

Box: exterior dimensions 46 x 46 x 41 cm (18 x 18 x 16 in). The interior compartment should have a dimension of 27 cm square and a depth of 32 cm (10½ in square, 13 in deep). At least 6 cm (2½ in) of peat is necessary all around to line the inner compartment. You should also leave a gap between the

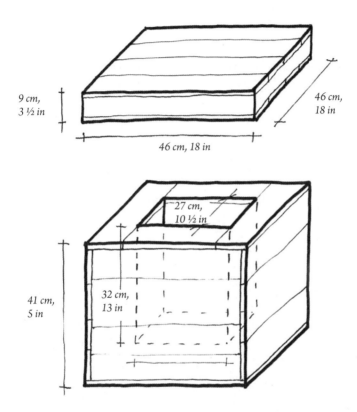

9 cm,
3 ½ in

46 cm,
18 in

46 cm, 18 in

27 cm,
10 ½ in

41 cm,
5 in

32 cm,
13 in

box and lid for aeration: insert a nail into the top of the box, sticking up 1 or 2 mm (½ in).

Materials: For the box and cover use slats (1.5–2 cm, ¾ in thick) of untreated pine, Douglas fir, larch or hardwoods (chestnut etc). The slats should be fitted tightly so they are watertight do not lose peat, which would diminish the quality of the preparation.

Using horn manure 500 and 500P

When to spray

Horn manure (500) preparation must be sprayed at least once a year, ideally twice – or more in specialist circumstances – normally once in spring and again in autumn, before the biological activity of the soil is at its most intense. Judge the requirements for your own land; these are general recommendations for land that gets enough heat and humidity.

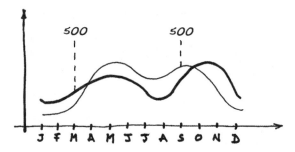

Graph of biological activity of the soil (in central France)

On the graph the fine line corresponds to the development of the soil's biological activity in average conditions in central France; the thick line refers to the coastal, non-irrigated regions of France. These indications may be of some help in adjusting for local conditions elsewhere.

These two sprayings are the minimum needed for the application to be effective. In some cases, for example when tending a market garden or during conversion periods, three or four sprayings can be done annually.

The spring spraying can be from the end of March to the beginning of May (depending on local conditions), when the ground begins to warm up. Earlier spraying is better, but if there's a tendency for frost days, don't intervene too soon as this can increase frost sensitivity in vines and trees.

The autumn spraying can be anytime from mid August to the end of October (again depending on local conditions), before the ground gets really cold. The earlier the better, as long as there is sufficient and consistent moisture in the soil. In warmer regions, start earlier in the spring and a little later in the autumn as long as there is sufficient humidity in the soil.

In pastures or meadows, spray in early spring and then, when needed, just after cutting or grazing, when harrowing the land, disking, finger tining or spreading manure. In autumn, apply as early as mid August if the humidity in the soil is sufficient.

On green cover crops, use the 500P at sowing time and after turning the crop under, which allows the green cover to be buried without risk of depriving the soil of oxygen.

For grain crops, spray just before or just after sowing. You can repeat spraying in the spring, which can be especially important in the case of weak vegetation. Spraying can also be beneficial when under-sowing with catch crops or green manure.

For market gardens and vegetable crops, it is best applied when sowing or transplanting the crop, in addition to the usual spring and autumn applications.

For vines, and fruit trees, spray just before or just after budding. (In cases of insect infestation, like cutworms or moths, the application of horn manure preparation can help regulation and accelerate budding, giving the buds protection from scavengers.) Take care when spraying early in springtime as this can make the plants sensitive to frost. In autumn, spray as soon as possible after the harvest. This preparation is essential when planting trees or vines; it must be sprayed when transplanting (the evening before or on the day), and also added to the paste in which the roots are wrapped. In the first few years after conversion to biodynamics, the spray can be applied intensively.

With a little experience, horn manure or prepared horn manure can be applied more often, particularly on irrigated areas with adequate organic matter (regular use of compost or cover crops). With vegetable crops, the 500 or 500P can be sprayed in the springtime and summer at intervals of several weeks. Use at every important sowing or transplanting time, since it favours the root system of the plants and development of microbial activity in the soil. The stirred horn manure is also used in seed or tuber baths, as well as being an important ingredient in root dip when transplanting fruit trees and vines. Horn manure spray is refreshing during heat waves or times of drought, and can be used on and around crops that need it, late in the evening or at nightfall.

Note that **overuse of this preparation** on crops (more than 3 or 4 passes on the same crop), especially as a foliar spray, can lead to poor storage qualities (such as premature rotting) or fungal diseases. This is also made worse if not enough horn silica (501), a necessary complement to the horn manure (500 or 500P), has been applied.

The reaction time to this preparation will vary initially from a couple of weeks to a couple of months, depending on history of the soil. Soils treated with pesticides or glyphosate-based herbicides may take up to two or three years before showing any effect from this preparation.

Horn manure isn't a replacement for regular manure, which can be vital for achieving good levels of productivity. However, horn manure stimulates the life of the soil and speeds up the formation of humus.

How to spray

Horn manure preparation (500) should be diluted in water and stirred energetically and continually for exactly one hour without interruption before spraying. Applying the preparation should be done in the evening, not before 4 or 5 pm. It is best on a day that is neither too hot nor too windy; the end of an afternoon that is partly cloudy is ideal. Avoid spraying during a downpour, but a little drizzle can be advantageous. Delay if an overnight frost is expected.

The right atmospheric conditions, and favourable soil condition, are more important than finding the right day in the biodynamic planting calendar. It is imperative though, to avoid the few hours during and after nodes and eclipses of the moon and planets. Choosing a day during a descending moon, particularly a root day during this period can impact positively, but is not essential (see chapter *Working with Cosmic Rhythms*). These periods are very brief (one or two 3-day periods in a lunar month) and waiting for those perfect conditions can be very stressful, or even impossible.

During drought periods and strong heat waves, spray the horn manure (500 or 500P) on vulnerable vegetation and around parcels of land. This should be done very late in the evening or even at night to minimise evaporation. Positive effects on the vegetation can already be observed the following morning. This can be repeated several times. Don't forget to also apply the horn silica preparation which works in conjunction with horn manure during the time before harvest.

Water quality

Use well stored rain water whenever available. A good method is to store water in a cement tank buried in the ground. The surfaces of the tank should be coated with tartaric acid, to prevent the limestone in the cement from leeching into the water, which increases the pH. It is also possible to obtain stainless steel tanks. Tanks with epoxy resin are not recommended, but are possible if the water is used within a short period.

After a long dry period it is best to allow the first rainfall to wash away before gathering rainfall from a roof. The roof should only be made of material such as slate or tiles, with steel, zinc and copper edging.

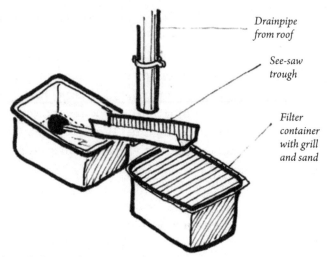

Drainpipe from roof

See-saw trough

Filter container with grill and sand

Position of trough showing first rainwater flowing into separate container

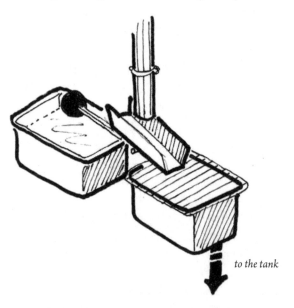

to the tank

You can construct a device to automatically separate the first rainfall from a roof, thereby avoiding collecting it in the rainwater tank (see diagram above). A trough or pipe made of copper or zinc, for example, see-sawing on a horizontal axis allows the first tainted rain to flow into a separate container. A ball-cock or floater is attached to the trough or pipe, and as the first container fills up, it tilts back into the rainwater tank. Allow 2 to 4 litres of water per square metre of roof (½–1 US gal/sq yd) to get a proper washing. The rainwater should be filtered through a container filled with sand, protected by a grill (to prevent it from being used by cats as a litter box). Remember that all rainwater is contaminated with synthetic pesticides, with the first few minutes of rainfall being the most polluted.

If not using rainwater, you can use water from a clean stream on the land or a spring if the water is not too hard (with a high mineral content).

Almost all tap water contains chlorine, nitrates and pesticides. Using this water greatly diminishes the efficacy of any biodynamic preparation, plant extracts or compost teas. If city water is the only water available, leave it outside in an open container for several days, stirring it briefly from time to time; but even then it is not ideal and can lead to poor results.

Well water, or water from a borehole, should also be enlivened before use by exposing it to the light for several days and stirring it a few seconds daily to help oxygenate it.

Check the pH and hardness of the water. Hard water, alkaline water or water with a high iron content are not suitable. Look for a pH lower than 6.5. Do not correct the pH of the water by adding something to acidify it. Try to get water with as little mineral content as possible and with a pH lower than 6.5.

Water temperature

The water should be lukewarm before stirring. Never add boiling water to heat up colder water, because water hotter than 37°C/98°F loses its ability to take on cosmic influences, which is what we want to achieve by stirring. Instead, heat all the water to body temperature (37°C/98°F). Use gas or wood to heat it, but avoid electricity or liquid fuel. Tip: aim for 35°C/95°F, so as not to exceed 37°C/98°F.

Different systems can be used: a burner or wood pellet stove, a water bath that will heat up to no more than 37 degrees, equipped with hot water circulation. Do not continue to heat while stirring, a gentle cooling of the liquid is preferred during the dynamisation.

Stirring (dynamisation)

Container

The container must be clean and uncontaminated (not previously used to hold any chemical product, even essential oils). It can be made of copper, stoneware pottery (without lead glazing), tin-plated or enamelled metal (avoid scratched or chipped surfaces as these can rust). If nothing else is available, you can use stainless steel. Rounded out bottoms allow for a better stirring effect. Wood is difficult to maintain. If unavoidable, the wood must be of good quality, and like a new barrel, be free from odours, and without mould or fungus, and must be thoroughly cleaned before each use

The shape of barrels and casks can affect the formation of the vortex. You can choose a container in the shape of an upside-down cone, such as a bucket; perfectly cylindrical shapes are, in fact, the best.

The container must be greater in height than diameter, the ideal ratio being 1.4 to 1. When filling, leave enough space to allow the formation of a good vortex when stirring.

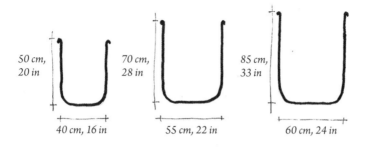

Mechanical stirrer as used in Australia, copper container

Quantity

Use 100 g of horn manure (500 or 500P) in 25 to 35 litres of water per hectare of land (1½ oz, 2.5–3.5 gal per acre). The quantity of the horn manure preparation for the area is critical. While it can be stirred in a larger volume of water, this is not economical for heating or convenient for handling. For rocky areas you might need 50 to 80 litres per hectare (5–8 gal per acre). For areas smaller than 1000 m^2 use 10–20 g in 5–10 litres of water (less than ¼ acre, ½ oz, 1½ –2½ gal) and spray it all.

Stirring method

Add the required quantity of preparation by crumbling it into lukewarm water and stir for exactly one hour (the method of stirring should ideally be learned from an experienced biodynamic practitioner). The stirring consists of stirring vigorously in one direction, forming a deep vortex, and then reversing the direction.

Form an energetic vortex without causing any water to spill over. The funnel created should be as straight as possible; there shouldn't be any shoulders forming. The vortex must be as deep as possible, going right to the bottom of the container.

The chaos must be vigorous. For a container of 100 l (25 gal), it will take 15 to 20 seconds to form a vortex which reaches the bottom of the container, before changing direction. For smaller containers, a couple of seconds is enough. Find a pace that is active and rhythmic, without any pauses.

Chaos during stirring

Create a straight funnel without a shoulder

Manual stirring lets you have time to get intimately connected to this work. Large areas or farms with specialty crops requiring several passes of the preparations may prefer stirring mechanically. (A number of companies make biodynamic stirring machines; contact your local biodynamic association for help in choosing one. Stirring machines developed in Australia are often made of copper, a strong material linked to revitalising forces, and have a sensor which tells the machine when to reverse the direction, creating a good dynamic rhythm. They can also be used for making teas.) It is not recommended to stir quantities greater than 120 litres (30 gal) manually or to stir quantities larger than 250 l (65 gal) mechanically.

Ideally only one person should stir for the entire hour, but it's possible to take turns as long as you concentrate on maintaining the continuity of stirring, and work without interruptions during the formation of the vortex and chaos. Stirring should be done in open air and light, and away from the shade of any high buildings. An area with good acoustics is favourable to the effectiveness of this preparation.

Manual stirring

Wooden hand-stirrer

Application

The horn manure preparation should be sprayed as quickly as possible after stirring is finished and filtering complete, and should ideally come into contact with bare soil, except on pasture, grasses or covered orchards and grassed vineyards.

Machine for heating, dynamisation, filtration and decanting [Andrea Kihlgren]

For pasture, it is good to spray after haying, grazing or cutting. In this case, one pass with a chain harrow (or similar tool), to loosely spread the manure and slightly open up the soil, allows the preparation to be more easily absorbed. In a market garden setting or orchard, the materials used to cover the soil (row cover, black plastic or mulch) should be removed. In a vineyard or orchard, the preparation should touch the soil as well as the plant.

Spraying should not be done at the same time as sowing or working the soil, because the preparation must be used according to particular rhythms. Taking short cuts leads to poor results in the development of the soil.

Filtration

Filter the liquid carefully through a fine sieve made of stainless steel, a linen, silk or cotton cloth, or (as a last resort) a nylon stocking. You can never be too careful: the smallest residual

Stainless steel filter

particle can block the sprayer nozzles and compromise the work, delaying the time between stirring and spraying.

Spraying

Spray immediately after stirring and filtering, ideally within an hour or at most two hours after stirring. Spray on the ground at a low pressure: 0.5–1 bar (7 to 14.5 psi) maximum, with medium droplets coming out, a consistent flow and as spread out as possible.

Copper backpack sprayer and various nozzles

For small garden areas, sprinkling with a small brush and pail is not ideal, but can be used when just getting started.

For medium sized areas (1–10 ha, 2–25 acres), for market gardens and vineyards, a copper backpack sprayer which has not been used for any other purpose (except for other biodynamic preparations, teas or decoctions) is ideal. If the tanks or containers used are synthetic, always rinse with pure water before and after use, store them in the shade and keep the cover slightly open when not in use.

For much larger areas, you will need to fix something on to a tractor. You can use old copper sprayers or heating tanks made of enamel. (These can be cleaned with a mild acid cleaner, scrubbed with vinegar with salt added, or soaked in a fermented nettle tea for a long time).

Do not use storage tanks or synthetic materials that have been in contact with pesticides.

Electric diaphragm pumps connected to the tractor battery are excellent. Attach a speed regulator and a manometer to control the pressure and flow. Piston pumps commonly found on agriculture sprayers are not good: the rhythms are too violent and jerked and the return to the tank generates a whole new stirring effect, which actually diminishes the effects of the preparations, especially horn manure (500 or 500P).

Quad bike with tanks and diaphragm pump

Turbine sprayer for horn silica (501) and large low-pressure jets for spraying horn manure (500 or 500P) with a rotating disperser (using a windscreen wiper)

Reinforced horn manure (500 Urtica)

Introduced by Volkmar Lust in Germany and developed in France by François Bouchet, this preparation consists of stirring the horn manure (500 or 500P) at 4 times its normal dose (400–480 g/ha, 6–7 oz/acre) in a nettle tea and leaving it to soak for 12 to 24 hours. It is sprayed directly onto the plants to allow foliar (leaf) absorption.

Prepare the nettle tea (100 g of dried nettle in 3.5 litres water for one hectare, 1½ oz in ½ gal per acre) and remove from the heat source as soon as it boils. After letting it sit for 12 to 24 hours, filter and dilute in nine times as much warm water (31½ l for 3½ l, 4½ gal for ½ gal) and add a quadruple dose of 500 (or 500P) and stir for exactly one hour.

Organic lettuce (left); biodynamic lettuce (right) which has been treated with prepared horn manure (500P) and horn silica (501) (trials by Adriano Zago, Fontanabona farm, Italy)

This reinforced preparation, applied to the foliage of plants, assists with some deficiencies or serious blockages in growth. It solves a good number of serious viral problems in viticulture. Where there is sufficient leaf surface area (at least four or five leaves), and until the end of flowering, it can be sprayed on vines to boost growth as needed. For the best results, ensure the leaves are well dampened by the spray.

Silica crystals

2. Horn Silica (501)

Horn silica (501) is made from rock crystal (the purest quartz possible), crushed into a powder with a little water added until it takes on a colloidal consistency; this is stuffed into a cow horn, buried and left to mature in the soil during the summer months.

It is an essential preparation for biodynamic agriculture, complementary and in polarity to horn manure (500 or 500P). It does not address soil issues *per se*, but works on the growing plant itself. In some ways, this preparation is an application of light that helps improve vitality during the vegetative (growth) stage but can also lead to excessive leaf growth.

This preparation brings a luminous, crystalline quality to plants and mitigates tendencies towards disease. It not only reinforces the effects of sunlight, it also gives plants a better relationship with the cosmos. This preparation is vital for the internal structure of plants and for their development. It favours the vertical growth of plants, and helps with the tying up of vines in viticulture. It strengthens the plants by giving them flexibility, and increases the quality and resistance of leaf and fruit epidermis.

It also ensures good nutritional quality, bringing out the taste and flavour of food. Storage is also improved. The use of this preparation is particularly important in greenhouses and low-sunlight environments as it compensates for the lack of light, as well as balancing the excess heat and humidity that favours fungal disease and hypertrophy.

In animal husbandry, it has been shown that the health of animals is improved through forage that has received the horn silica preparation, and their milk and meat are likewise improved.

Storage

The preparation is stored in a glass jar, covered but not airtight. The ideal is a jar with a glass lid, with the metal locking frame removed.

Store this preparation where there is some sunlight (preferably where it catches the morning sun), but avoid too much direct sunlight. A ledge facing east or northeast is ideal. Avoid storing near any power lines or cables, or where there is electro-magnetic pollution from wi-fi, mobile cell phones, etc. If you can't find a suitable place in a building, install a small, elevated wooden platform in a hedge or bush, to hold the jar (make sure it has a rim, to avoid accidental falls).

The contents of the jar should be shaken from time to time. Stored in this way, the preparation should last for several years.

Storage of horn silica preparation

Using horn silica

When to spray

The use of horn silica preparation can start after the horn manure preparation (500 or 500P) has been applied. As horn silica preparation works on the exposed parts of the plant, it is important to consider the developmental stage of the plant, as well as the weather, when spraying (see chapter *Working with Cosmic Rhythms*).

Spraying horn silica (501)

Manual spraying with adapted hoses

Mechanical spraying of horn silica (501)

The 501 preparation is applied most frequently in the spring and in the autumn. It is particularly effective after heavy rainfall or during prolonged periods of humidity, but avoid spraying during a downpour.

In the spring, an early application is often the most effective for the growth and development of the plant. For legumes and grain, however, it must always be applied when there is no chance of frost. It can be applied several times throughout the spring and in the summer if there is sufficient irrigation or humidity.

The optimal effects are obtained by spraying when the plant is at its strongest growth; for example, on grain crops when they are tillering (producing additional stems at the base of the plant), during stem elongation, and during formation of the seed head. Horn silica spray on grain crops has a tendency to increase the protein content of the grain.

On fruit trees, apply 501 once the leaves start to really grow, early to mid-spring, and once the fruit is formed but still quite small. Don't apply during the flowering period. Spraying one more time once the fruit is completely developed helps the fruit to ripen.

On vines, one to three applications are possible in the spring, beginning at the five leaf stage, and ending before flowering commences. If it is a rainy summer and risk of fungal disease is high, additional spraying can be done from setting until *veraison* (the onset of ripening). Spraying one to three weeks before harvest (the weather being neither too hot nor dry) greatly increase the quality and rot-resistance of the fruit.

For pastures, it is best to wait until the grass is taller before spraying (young shoots of at least 10 cm, 4 in). As a general rule, wait 8 to 10 days after doing a pass of the horn manure (500 or 500P).

On lettuce and cabbage, especially when bare root transplanting, avoid the horn silica spray until after transplanting is done. It is best to wait until the roots begin to set, and the first signs of new growth and leaf roll are seen; if not, there is the risk of the plant going to seed.

As a general rule, try to **avoid spraying 501 on anything that is too young or too weak. Never spray 501 on plants suffering from drought or dryness** because the 501 encourages the plant to sweat and lose even more water, potentially causing considerable damage.

For plants in pots or in raised beds, and particularly in greenhouses, spray before transplanting if the need arises (for instance, plants that get leggy due to lack of light, or plants showing tendencies towards fungal disease).

How to spray

Apply 501 immediately after stirring is finished. The best time to spray is shortly after sunrise (not before) and preferably before 8 am. It is very beneficial if the morning dew is still out when spraying, and to choose a nice sunny morning with little or no wind. The temperature must not be too hot when spraying (maximum 22°C, 72°F).

Spraying on cloudy or misty mornings is possible if there is a long period of humidity or lack of sun. You can spray onto damp leaves; don't spray when its raining, but a sudden shower soon after spraying won't lessen the effect.

There are especially blessed days that are hard to describe in a manual: you have to experience them and feel the moment.

Evening spraying

There are some exceptions to the above. For root vegetables like carrots, beets, rutabaga (swede), and celeriac, 501 can be sprayed in the evenings, two to three weeks before harvesting. This practice increases sugar content, and improves the taste and storage quality of the vegetables. This evening spraying can also be done on vines after the harvest, and before the leaves fall off. It reinforces the maturing of wood and sends energy into the roots. This process allows the buds and young shoots to unfold more regularly in the spring, and facilitates a better start-up of growth. However, this should only be necessary if there is a significant lack of woody growth or if spraying hasn't been possible during the usual time due to climate or vegetative weakness.

Some practitioners recommend this evening spray on fruit trees before the harvest. This is favourable to the mature fruits, but own observations have been that this practice devitalises the tree.

Stirring (dynamisation)

The basic method is as for horn manure (see p. 34), but it can be slightly modified. For horn silica (501) the quality and temperature of the water have less of an effect. It is preferable not to stir at night, especially if using a stirring machine; wait until the very first signs of dawn. Heating up the water, however, can be done beforehand.

Manual stirring for small quantities

Quantity

Use 2 to 4 g per hectare (depending on quality of the silica preparation) in 25 to 35 litres of good quality water (½ oz per 10 acres with 30–40 gal). In certain cases where the leaves are over-developed, as in arboriculture, one may need to wet the leaves more. In this case, use 50 to 100 l/ha (50–100 gal/10 acres) instead. For a field of 1000 m^2 (¼ acre) or less, stir ¼ g into 5–10 l (5–10 quarts) of water for one hour and spray the entire amount.

Stirring method

Stir energetically for exactly one hour, just as in the horn manure preparation.

Bucket and stirring stick as used by the author

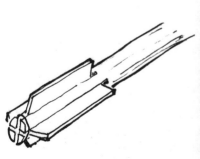

Detail of stirring stick

Wooden barrel for manual stirring of larger quantities

Application

Filtration

Despite the fineness of the silica, it is best to filter before spraying as some particles may clog up the nozzles during spraying.

Spraying

Spray immediately after stirring and filtering, at most 3 hours after completing the stirring. This preparation should be applied with high pressure, as a fine mist pointing upwards to allow it to fall down onto the leaves. Choose a pressure of at least 2 to 3 bars (30–45 psi); you can go up to 12 bars. The aim is not to wet the leaves but to create a fine mist for the light to penetrate. Keep this in mind when observing where the preparation falls after misting out. This fineness is more difficult to achieve when there is a breeze; in the early morning, the wind is usually calmer.

Double filter used before pouring preparation into spray tanks

Spraying a fine mist over a vineyard

About horn silica (501)

Horn silica is a powerful element and essential for all biodynamic crops (Demeter standards require it be done at least once on all crops), but it must be used with caution. Plants that are too weak or too young should not be sprayed. Never spray on plants that have stopped growing due to cold, dryness, or heat wave. When growth is poor because it has been sprayed mistakenly on plant that was too weak, for example, you can remedy the effects by spraying horn manure (500 or 500P) on the foliage in the normal dosage.

In certain cases, spraying on plants too late in the morning can cause damage to the foliage (causing the plant to burn). In our experience, three conditions can cause this type of damage: repetitive spraying with too little time in between (for example, a few hours); weather conditions that are too hot or dry; and spraying on plants that are too weak, or too young and small (if necessary, cover them up during spraying). During very hot conditions, to be safe, spray before sunrise. Horn silica (501) causes the plant to sweat so a good watering on the days following application can be beneficial, especially during a heat wave.

If 501 is used after a dry period and there is little irrigation available, it may bring the plant too much light, more than it can handle, thus forcing the plant to go to seed, or causing it to dry out.

Avoid spraying certain flowering legumes as well as flowering vines or orchard trees, since the flowers will have a tendency to dry out. There are, of course, many exceptions, in particular strawberries, tomatoes and the squash family; these can all receive this preparation without ill effects during the flowering period.

Note about spraying the preparations

Learn to look at the soil and plants, to observe their tendency and behaviour, and to plan your spray times for the 500 and 501 accordingly.

All steps and procedures concerning the quantity and times for spraying are only valid for preparations that are of good quality, of colloidal nature and stored in the best possible conditions. The preparations must be well stirred and applied with the correct pressure.

With **horn silica** take more care in choosing the right time to spray.

However, for **horn manure** the storage conditions, the stirring and spraying techniques are more crucial, as its organic matter is a living substance that is more fragile. The utmost care is needed when working with the horn manure: the quality and temperature of the water, the rhythm of the stirring, as well as the need to use it immediately after stirring.

Here too, you can see the polarity of these two preparations, working to complement each other. The organic 500 or 500P is very delicate and sensitive to how and when it is used, while the mineral 501 is stable and less demanding.

Effectiveness

You can primarily judge the effectiveness of the preparations by observing changes to the soil structure. It should become more crumbly, increasing its porousness, and it should appear darker with a more refined scent. It should seem more alive.

Strong root formation – both deep and lateral – is another good sign, with root secretions directly influencing the rhizosphere, as is the increased ability of the soil to absorb and retain water: you will notice the soil feeling softer and bouncier underfoot.

Pay attention also to the appearance of all flora and the physical characteristics of plants, such as the flexibility, position, colour and luminosity of the leaves. Finally, increased resistance to disease, improved smell and taste of the harvest, and good storage of the harvest are all signs of the effectiveness of horn manure and horn silica.

Biodynamic compost piles

3. Biodynamic Compost

The care and composting of organic matter is fundamental to biodynamic agriculture. Biodynamic compost is characteristically made by composting in rows or piles and covering with straw, old hay, earth, etc., before specific preparations are introduced.

These preparations not only act on the compost itself, but in fact work especially on the soil where they are spread. It's not that they introduce elements to the soil, but rather than they mobilise elements found there, and in the cosmic surrounds. The preparations act as a leavening agent, introducing vitality and a new-found health to the soil, through the compost.

Semi-paved compost area built using the Pfeiffer method (controlled fermentation with heat and air) (Rengoldshausen Farm, Germany)

The right site for a compost pile

The site should be flat, or better slightly domed, so as not to collect water; the base of the compost pile should not be waterlogged. About 2–3 metres squared (6–9 feet squared) are needed per cubic metre (35 cubic feet) of raw compost.

The pile should not be too exposed to wind, or summer sun, but surrounded and shaded by trees and shrubs (elderberry, birch, alders, hazelnut, rosehip, willow, dogwood, wild plum, types of *spiraea*, etc), without being too close to trees roots which might draw energy from it. An elongated pile is best oriented north/south to better spreads the sun and heat.

Avoid building the pile under a shelter or on concrete, which prevents micro-organisms, bacteria or worms getting into the pile. In case of legal restrictions, for instance, in a water catchment area, some biodynamic practitioners (for example, at Rengoldshausen Farm in Germany; see illustration above) have solved this problem by using bioterre bricks for a base that allowed them to catch all the run-off from the compost – which was then used to make a tea or to spray onto the compost when dry.

Preparing the soil

The grass turf should be cut and turned over, preventing rotting. The ground could eventually be worked over very superficially.

If the ground has never been worked, spray the 500P or the barrel preparation, in the normal way. It is important to maintain the same spot for compost piles, as fungi, micro-organisms and red worms establish themselves there and recolonise a new compost pile quite rapidly.

When to make the pile

Ideally start in the spring between March to May, preferably during the descending moon. Build the compost using vegetable scraps, accumulated manure piles from the winter or fresh barn manure. This encourages a quick breakdown process and the compost pile can then be used in summer or autumn in most cases. One can also start composting at the end of the summer or beginning of autumn, especially in hot climates and if there is no way to irrigate the compost pile during summer. Autumn composting is useful if young compost is needed in the spring, or well-matured compost for the following autumn which can then be used for light soil fields or delicate crops. Composting dead leaves for the manufacture of potting soil is begun in the autumn and takes one or two years.

Choice of materials

Cow manure is the best material as a base. Obtain as much as possible, and try to get the best quality – nothing replaces the unique quality of cow manure, which is determined by:

⚘ The nature and origin of the straw: wheat straw is best; chemical treatments like fungicides must not be used, nor a straw chopper.

Good quality cow manure has a spiral form

- ψ The quantity of straw: the more straw there is in the manure, the better start the compost pile will have.
- ψ The state of the dung in relation to cows' nutrition, the general conditions they are raised in, the type of stalls used, their productivity, and of course their overall health.

There needs to be a good ratio of carbon to nitrogen. This C/N ratio should be around 30; that is, the material should contain 30 times more carbon than nitrogen. This is generally the case for composts made up of manure well embedded in straw from the barn. You need about 5 to 8 kg (11 to 18 lb) of straw per unit of livestock.

The pile can also be complemented with other manures that each bring their own special quality for different soils and crops. The 'hot' manures – for example chicken, horse, sheep or goat manure – are good for clay soil, which tends to be heavy and cold. 'Cold' manure – like that of cow, duck or pig – is best used on dry, sandy soil or where it is too sunny.

Almost all organic substances, whether vegetable or animal, can be composted on their own or mixed. You have to balance the pile between carbon, nitrogen and calcium. Remember that carbon is present in all mature vegetable matter (straw, old hay, dead leaves, cut branches, vine trimmings, debris from the forest, etc). Nitrogen is present in all animal matter (droppings, feathers, hair and silk, blood, offal) as well as in young or green vegetable matter like garden waste, grass clippings and green cover crop cuts.

You can also have a vegetable compost pile made of straw, old hay, hedge clippings, etc. These materials are best presoaked in rain water or, better, animal manure, and complemented with nitrogen-rich substances like pulverised feathers, horn, etc. Household scraps and garden and vegetable waste are often too wet and deficient in carbon; add straw or old hay to improve the carbon/nitrogen ratio. Allow fresh vegetable material to wilt before composting.

Allow fresh vegetable matter to wilt before composting

Other materials like rock powder, pig bristles, chicken feathers, uncauterised horns, and maerl, should be used with caution, and only when really necessary. If possible consult with an experienced farmer.

60

Dead leaves should be composted separately, possibly adding a little manure or starter to improve the carbon/nitrogen ratio and to help in the rotting process. Adding calcium brings an element of animal origin (most calcium or limestone is ancient animal remains). This compost is valuable in the garden and orchard, and can be used to regulate the vitality of trees that are too prolific.

Be very wary of materials collected outside of the farm. Find out if pesticides or other chemicals were used, especially in straw and manure from a conventional farm. Almost all straw today is treated with anti-fungal, anti-yeast and anti-bacterial agents.

From a biodynamic (or Rudolf Steiner's) point of view, human waste (from dry toilets, for example) shouldn't be used on soil where human food-chain plants are growing. However, it can be useful for open spaces, forests or pastures for animals which aren't eaten directly by human beings (dairy cows, for example).

Mixing the compost pile

Your own experiences, and trial and error, are central here. The more fresh and rich the manure is, the more straw and vegetable matter rich in carbon will be needed. Adding earth prevents excessive temperatures, especially in horse manure. Adding mature compost before the finishing layer is an excellent practice. You can add up to 10% of the total volume.

If the compost is only made of vegetable matter, adding some calcium in the form of slaked lime, ground limestone, dolomite or even maerl can help.

Cut fruit-tree branches, vine trimmings, hedge clippings, etc, (preferably chopped up or crushed) and grape skins need to be pre-composted for a year, adding materials rich in nitrogen. Reintroduce in the following spring into the final mix of materials for the compost. In a few months this mixture will have totally composted and be ready for autumn.

Muck spreader with a deflector, for making small compost piles

A well mixed pile makes for better results. A compost turner and muck spreader are useful for crumbling material that is too compact and can help to build uniform piles. Putting a deflector at the back of the muck spreader gives a uniform pile and limits its height. However, the use of the spreader or compost turner can introduce too much air, especially in horse manure. In this case, use the front loading forks of a tractor and let the material slowly fall down. This will disperse it better with less overheating, and thus less loss. Often the borders will need to be attended to manually to give a regular shape to the pile.

Moisture

A good level of moisture is important from the start. To check, take a handful of compost and squeeze it. It should not drip, but you should feel some liquid between your fingers. If necessary, set up a sprinkler system, for example, a micro sprinkler spraying

20 litres per hour, over a diameter of about 1.8 metres (5 gal over 6 ft diameter). Sometimes more than one day's watering may be needed. Check the compost carefully to ensure it is not overwatered. It is often best to do this in several stages, allowing the water to penetrate through the whole mass of the pile.

In hot, dry climates where it is windy, it may be necessary to use a permanent sprinkler system during a hot spell to maintain the humidity of the pile.

Making the pile

For small quantities, make a central core with layers of straw from a square bale (medium density). For larger quantities, use old dry straw or unravel a round bale, especially if the material (manure, etc) is too wet or compacted. Alternatively, form the pile on ground that has been cleaned of any fresh vegetable matter, and levelled off.

The pile should not be higher than 1.3–1.5 m (4–5 ft) in height, and 1.5–2 m (5–7 ft) in length.

To finish the pile, ensure that the sides slope gradually (but steeply enough that rain water runs off), leaving 20–60 cm (8–24 in) width at the top. The top should be domed or rounded off in the winter, and indented in the summer – this allows, respectively, for water to run off or collect.

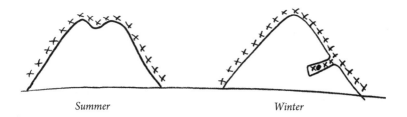

Summer   Winter

The biodynamic compost preparations

Originally, the description given by Rudolf Steiner of the activity of the biodynamic compost preparations was not actually about the process of composting itself, but about their effects on the soil and the behaviour of the plants.

However, experience has shown that the preparations are imbued with interesting properties such as limiting overheating, reducing loss of substance, and improving conservation of nitrates and phosphates.

Contrary to what some people think, the preparations are not really agents of disintegration, but rather help to organise and structure substances.

The six preparations all come from the plant kingdom, although some are first fermented in animal material. They are referred to by the name of the plant, or simply by the numbers 502 to 507.

Yarrow (502)

Yarrow (*Achillea millefolium*) plays a significant role in the processes connected with sulphur and potassium, and a secondary role regulating selenium and silica

Camomile (503)

Camomile (*Matricaria recutita*) is connected to the metabolism of calcium, and regulates the nitrogen process through reduced loss of ammonia; it has a secondary role regulating potassium, boron and manganese

Nettle (504)

Nettle (*Urtica dioïca*) relates both to nitrogen and iron; it reinforces the influence of the two first preparations. It gives both compost and soil a sensitivity – a sort of 'reason' – and encourages the breakdown of organic matter into humus. It has a secondary role regulating potassium, sulphur, calcium, magnesium and manganese.

Oak Bark (505)

Oak Bark (*Quercus robur*) has a special relationship to calcium and mitigates plant diseases due to its prolific, exuberant character. It has a secondary role regulating phosphorus.

Dandelion (506)

Dandelion (*Taraxacum officinalis*) helps to regulate silicic acid and hydrogen. It also guides the developmental processes of potassium and limestone and ultimately those of nitrogen. It also has a secondary role regulating boron, magnesium and selenium.

Valerian (507)

Valerian (*Valeriana officinalis*) is a liquid preparation which stimulates the phosphorus processes in the soil and forms a sort of warm, protective blanket around the compost, like the skin of an organism. It also has a secondary role regulating magnesium and selenium.

The efficacy of these preparations is largely dependent on the care taken in making them. The timing and method of harvesting the plants, and of drying and storing them, is very precise. The quality of the animal parts used in the preparation is also important, and is directly linked to a respect for the nature of the animals.

The best way to make the preparations is to learn the

techniques from an experienced practitioner; indeed, on a farm, the knowledge of those who prepare, care for and conserve the preparations in the key to their effectiveness. Alternatively you can often buy them from your local biodynamic association. Make sure you order the preparations in good time, and work out the quantity you will need. When you receive them, they can be stored for a few days only by wrapping them in balls of old compost.

Storage of the compost preparations

Like the horn manure preparation (500), the compost preparations must be carefully stored to avoid any reduction in their efficacy. They must be stored individually; you can build a box with six compartments suitable for these preparations. The size should be in proportion to the containers used. It is not good to store small containers in a large box.

See the section on storage of horn manure (p. 22) for basic instructions on making a storage box and insulating it with peat. The diagram below gives sizes for a storage box suitable for 10 to 100 ha (25–250 acres) or more.

Regularly check the humidity of the preparations. Even in a good storage box they can dry out easily, which will alter their effectiveness.

Storage box for the preparations

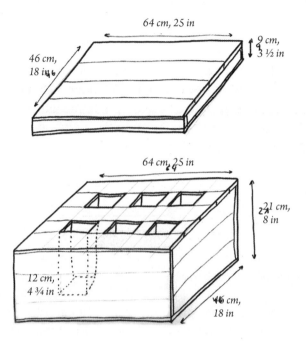

Quantity to use

Every 10 to 15 m³ (350–500 cu ft) of compost requires 2 g (about 1/20 oz) of each solid preparation and 2–5 ml (½–1 tsp) of valerian (depending on the quality – be guided by the smell). For the valerian, the quantity of water used to stir with will depend on the size of the pile: for compost volumes of 5 to 50 m³, one to two litres will suffice. For larger piles, adjust the amount of valerian and rain water proportionally; for example, for 500 m³, use 250 ml valerian in 15–20 l of tepid rain water (17,500 cu ft, 8 ½ fl oz, 4–5 gal). Stir for 10 to 20 minutes.

If the farm has cattle, as a guide you need 1 to 2 sets of preparations per livestock unit. For a market garden, you need 10–12 sets per hectare (4–5 sets per acre), and for diversified farms, use up to 2 sets per ha (up to 1 set per acre).

Inserting the preparations into the pile

Ideally, insert the preparations on the day you build the pile (so have the preparations ready). A pile 1.5 metres (5 ft) high and 1.75 m (6 ft) deep contains 1.3 m³ per metre length (14 cu ft per ft length). Every 10 to 15 m³ (350–500 cu ft) needs one complete set of preparations.

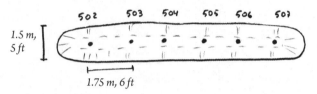

Vertical holes along the top of a pile

The preparations will be inserted at equal intervals from each other, about 40 to 50 cm (16–20 in) deep. For a pile that is 8 to 10 m long (25–35 ft) make 6 holes vertically along the top about 1.5 m (5 ft) apart. Alternatively, the holes can be made on each side of the pile at an angle of 45°, about 3 m (10 ft) apart. Use a stick of about 6–8 cm (3 in) diameter to shape the holes big enough (10–12 cm, 4–5 in diameter), to a depth of about an arm's length.

One preparation is placed in each hole. Wrap the preparations in compost or good soil, making a ball roughly the size of a plum or golf ball. Ideally, the material used to form this ball should be of the consistency of the horn manure preparation. Place them in gently and, if you can, place a handful of compost or good garden soil at the bottom of the holes.

The 507 preparation is a liquid extract of valerian. Stir this in lukewarm rain water in the same way as for horn manure preparation, but only for 10 to 20 minutes. Pour about 20 ml (1½ tbsp) of liquid into its hole, reserving the rest for spraying (see below).

Close the holes with old compost or good soil ensuring you don't make any depressions that might collect water.

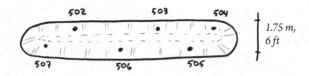

Angled holes along the sides of a pile

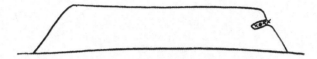

Side view (top) and plan (below) of a small pile

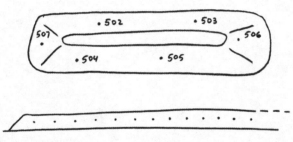

Side view of a long pile

Spray the top of the pile with the remaining valerian, giving it a protective coating. The spray should be very fine and even, preferably made with a copper backpack sprayer. If using a small brush and pail to sprinkle the preparation, use more liquid to achieve the same blanketing effect.

Covering the pile

A compost pile is a live organism that needs a skin. Make this from layers of straw from a square bale or fluff up straw from a round bale; or use old hay, plant cuttings, soil, etc. The covering should be at least 15–20 cm (6–8 in) thick and well made to

A well made compost pile

A compost pile with a permeable covering

Compost heap carefully covered with straw

ensure success. If using soil as a covering, do not use too much as this prevents the pile from breathing; 2–5 cm (1–2 in) thick is sufficient.

In the winter in rainy regions, extra protection against heavy rain may be needed – but a little bit of rain is good for the compost pile.

If none of these materials are available, you will need to use a tarpaulin which is breathable and waterproof. For instance, TenCate Top Tex, a fabric made for organic compost piles, is suitable. Avoid tarpaulins made of plastic or synthetic films which are too watertight. (If you do have to use synthetics, ensure that there is a gap at the base of the pile to allow for breathing.)

Turning the pile

For well built compost, turning it over is rarely necessary. Generally speaking, only specialists should turn over the compost pile. If the compost pile has not developed well – for example, it is too wet, or too dry – then turning might be recommended along with other measures (adding more dry material, or spraying, as appropriate).

If really necessary, turn the compost pile slowly, and avoid mixing snow or frozen matter into it. Turning material that is too wet smooths the particles, preventing good breathing. Be careful with compost that has been turned over too many times or has been overly aerated: its apparent maturity does not reflect its true state. It can cause problems in the soil, similar to those created by nitrous materials.

A good compost pile

In summary, a well balanced compost pile is small with good quality diverse materials. The quantity and quality of straw determine how well the pile is aerated. It should be humid, but not excessively damp. The carbon/nitrogen ratio should be around 30. The covering of the pile must be adequately thick (if possible 20 cm, 8 in, of straw). It is best to work with the seasons, starting the pile in spring or at the beginning of autumn. And, of course, it must have the biodynamic preparations added, ideally on the day the pile is formed.

When to use it

Times to spread it

For most purposes, autumn is the best period to spread compost, from the descending moon in August to the end of November. This would be after the harvest and before ploughing or disking the field. For green cover crops this would be before planting or before turning over; on pasture before regrowth in the autumn or after cutting; and in orchards and vineyards shortly after harvesting. It is said that spreading compost on grazing land after mid August does not affect the rate at which animals eat the grass.

When is it mature?

Generally it is best to use the compost when it is almost (85%) decomposed. Look for a consistency that is crumbly, homogenous and colloidal. You should not be able to recognise any materials that have gone in to make the pile. As a test, roll a piece of straw or hay that still has its shape between your fingers. If it immediately starts to break up and turns black and colloidal, the compost is at a good stage. Good compost should not dirty the hands. It should smell like good earth or the undergrowth, without any hints of ammonia, mould or burnt smells. The compost worms (*Eisenia foetida*) should still be active. The length of the process varies according to the seasons, but is normally about three to six months. If the compost looks like black potting soil, and is as dry as the forest floor, it is too old and has lost its essential vitality; it will no longer be able to stimulate soil life.

There are exceptions, however. Certain crops may require different spreading times, and younger compost that has just finished heating up may also be used differently. For instance:

- ⚘ On a field crop in the spring, to bring a fertilising effect on the winter crops (liquid manure treated with the preparations would act in the same way).
- ⚘ In a market garden, on primary crops and varieties that tend to be insatiable such as tomatoes, eggplant, celery, leeks, cucumber family, etc.
- ⚘ In the orchard, in the spring, for varieties and species with early setting fruit like cherry and apricot.

Other exceptions have to do with the type of soil. It is best to use young compost in spring on heavy and cold soils with slow mineralisation; on soil which is blocked due to excess calcium; or on soil where an excess of clay has slowed down its metabolism.

In certain cases, for instance with delicate crops sensitive to disease, very vigorous plantings, or on soil that is very cold or with little microbial life, very mature compost – almost to the point of looking like potting soil – can be used with benefit.

If in doubt, consult an experienced biodynamic practitioner.

Sheet mulching

On field crops and pastures, one can use manure from the barn treated with the preparations direct (prepared manure), without putting it in piles and allowing it to compost. For large crops, it is important not to bury the prepared manure too deeply in the soil. Spring-time harrowing or finger tining superficially (5–8 cm, 2–3 in, maximum) should suffice.

There are some situations where spreading prepared manure on living soil, followed by spraying horn manure (500 or 500P), is beneficial, such as on green cover crops (during growth, or before mowing) or on hardy spring crops such as corn or forage crops like sorghum.

Care must be taken with soil that is not living, as fresh, non-composted manure contains bacterial elements which can be harmful to plants and the life of the soil. It is during maturation (usually over 3 to 6 months) that good compost develops a diverse flora of beneficial micro-organisms which can help control a range of plant diseases.

Jacques Fuchs of FiBL in Switzerland has written:

> Compost acts both directly and indirectly on the health of plants. Indirectly, it influences the structure of the soil and brings nutritional balance, especially trace elements. Most important, though, is its direct influence which comes from the beneficial microflora it contains.

Stages of compost

Contrary to the practices of conventional or organic composting, the heart of the pile should not get hotter than 55°C (130°F). If the temperature rises beyond this, spray the top of the pile. (The disinfection of compost matter takes place through organic processes at a temperatures close to 50°C, 120°F, over a period of more than three weeks.) In rare cases where the temperature of the compost does not rise, you have to turn it to get air into the pile and to rid it of excess humidity (see instructions above for turning a pile). Use a probing thermometer to check the temperature inside the pile. (You can make a simple tool with an ordinary thermometer and a sturdy 1 m/3 ft long tube made of PVC sheathing used for electrical installations.) Never leave the inside of the compost pile exposed either to air or to the sun when turning it or using it. Begin at one end and work along the length of the pile. When spreading compost, do not leave it on the surface (unless you're using a protective plant cover, such as on pastures, or when grassing orchards or vineyards in the autumn). To protect it, it should be harrowed into the soil to a depth of 5–8 cm (2–3 in).

The compost normally goes through four stages that give it its necessary qualities.

1. Thermal phase
2. Fungal and bacterial phase
3. Development of micro and macrofauna (woodlouse, springtail/Collembola, etc...)
4. Compost worm phase

These are broad suggestions. Experience and practice will bring a more detailed knowledge. **Composting is an art:** much like working with horn manure (500 or 500P) and horn silica (501), you need to develop and nurture a personal connection and commitment to the manure and compost.

Always remember it is better to use small quantities of good quality compost that has an earthy and colloidal nature, than to use tons of fresh organic matter or mediocre compost. The use of compost is not only a relationship between humic substances and mineral elements; it is also a stimulant for the life of the soil and the plants, an enlivening agent and regulator.

Remarks about compost

Very poor soil, that is soil which is light and deficient in organic matter and clay, needs to be transformed into more stable humus. In such a case, add 20 kg of powdered rock calcium or calcium phosphate, or even 10 kg of limestone or sifted wood ash, per tonne of compost (45 lb, 22 lb per ton) to encourage the process.

The addition of a large amount of clay in the compost pile works in the same way. Use 20–30 kg of dry clay per tonne (45–65 lb/ton), well spread so as not to choke the pile.

Another tip is to regularly spread a layer of non-soluble minerals like marl, natural phosphates, basalt or maerl onto animal bedding. This, in addition to the disinfection of the animal bedding and a good ambiance in the barn, will help provide good quality material to be composted, which in turn will encourage more stable humus.

Wooden bin for kitchen waste

4. Compost Preparations, Barrel Preparation and Starters

There are many variants of compost preparations. In Germany, many farms make their own according to a particular recipe, adapted for their land and the specific animal manure available on the farm. This chapter will discuss the first compost preparation, the birch pit preparation developed by Max Karl Schwartz; Maria Thun's barrel preparation, developed from work by Ehrenfried Pfeiffer and noted for its ability to counteract the effects of radiation; and compost manure with nettle, developed by American Walter Goldstein, which also gives good results.

These compost preparations are fairly weak, so they need to be sprayed several times for good effect. However, they're quite straightforward to make on a farm, or by a group of gardeners. They keep well in less-than-ideal conditions, because they're more like compost than a true biodynamic preparation. They're less sensitive to the quality of the water used, and the quality of the spraying equipment. They also keep for several days after stirring.

Their use varies in different parts of the world; in New Zealand and India, for example, Peter Proctor uses them at the very important dose of 2.5 kg/ha (2 lb/acre).

Maria Thun's barrel preparation

Maria Thun's barrel preparation has proved to be an excellent decomposer of organic substances and is easy to make in large quantities on a farm, or as a group working together. It can be used over large areas or on speciality crops when there is no prepared biodynamic compost; however it isn't a replacement for manure. It can be sprayed a few times before using horn manure (500).

Another approach is to use prepared horn manure (500P) instead of several applications of barrel preparation or simple horn manure (500). One application will bring the benefits of good compost (its various actions on organic matter, stimulating the elements in the soil and its positive effects on plant health), as well as those of horn manure (structuring of the soil, development of humus and working on the root systems).

Regular use of barrel preparation is justified in certain cases, for example on heavy calcareous (chalky) soil, where the organic matter is stagnant or blocked, or when a large amount of organic matter needs to be broken down, such as after a harvest or when turning under green cover crops. It can also be used before composting animal bedding, and to help mature and remove odour from ferments and liquid manure.

Maria Thun's barrel preparation in half buried barrels

Method

Based on advice from Maria Thun in 2011: collect 5 pails (about 50 kg, 110 lb) of fresh cow manure preferably from pregnant cows. Collect one or two-day-old, well formed cow-pats from the pasture. It is best if the pats have dried a little. Find a clean surface, cement or wood, to mix it. You can even use a cement mixer for this.

Mix the manure with 500 g (1 lb) coarse basalt and 100 g (3–4 oz) of finely crushed eggshells. One, or preferably two people, should mix this for an hour, turning it over, cutting it and repeating, as if making cement.

Take a barrel without a bottom (an old wine barrel of 110 l, 30 gal, for example) and bury it halfway in the ground. Alternatively line a pit with slats of wood or small logs (preferably white wood, like birch, elm, ash, poplar).

Fill the barrel or the pit with half the manure. Make five holes about half the depth of the manure. Add a pinch (2 g) of each compost preparation into the holes. The nettle preparation can be placed in the middle. Close the holes. Fill the barrel or pit with the other half of the manure and again

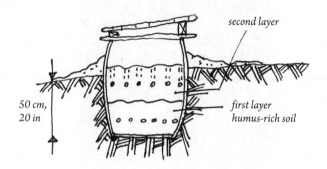

Maria Thun's barrel preparation

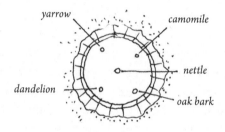

make five holes and add the five preparations. After closing the holes, sprinkle valerian preparation (507), prepared as described earlier (2–5 ml, ½–1 tsp, of valerian stirred for 10 to 20 minutes in 500 ml litres (17 US fl oz) of warm water) on the compost and around the barrel.

After one month, take out the contents of the barrel or pit and mix carefully for several minutes. Then put it back in the barrel and again add the preparations in two stages, as before. After this, the compost must mature for another 1–5 months, depending on the season and the initial quality of the manure. Once mature, when it's black and smells like good earth, it's ready to use.

Storage

The mature barrel preparation is often stored in the barrel or pit itself; this is easiest if using it right away. However, it is better to store it like the horn manure in a stoneware pot, glass or ceramic jar, stored in a wooden case lined with peat (see p. 23 for a description).

Application

The barrel preparation is used at a dose of 240 g/ha, in 35–50 l of water (3½ oz/acre in 4–5 gal), lukewarm if possible. Stir it vigorously in one direction and then in the other, for 20 minutes. Once stirred, the liquid keeps its efficacy for at least 72 hours.

Spraying three times at regular intervals gives good results. This can be several hours apart, or three evenings in a row, or even during a trine, following the guidelines in *The Maria Thun Biodynamic Calendar.*

Use the same principles when using it on animal bedding as a starter for composting. Adjust the quantity of water according to the dryness of the bedding, minimum 240 g in 15 l (8 oz in 4 gal) of water for a barn of about 50 cows. Spraying can be done with a brush, a backpack sprayer or even a sprinkler. The operation can be repeated at a various times (weekly or monthly) depending on the results. For more on its use with animal bedding or as liquid manure, see chapter *Liquid Manure and Animal Bedding.*

Birch pit preparation

Birch pit preparation is particularly recommended for poor, acidic soil. It is used mostly in the barn like an activator for animal bedding. It can be spread on animal bedding or before removing manure from the barn. It is simply mixed with warm water; stirring is not necessary but it doesn't hurt to do it. It was studied and used in the 1930s by Max Karl Schwartz.

Method

Dig a pit 60 cm (2 ft) wide and 40 cm (16 in) deep. The length is dependent on use. If the bottom of the pit is below arable soil, add a layer of good earth, 5–10 cm (2–4 in) deep. Avoid digging the pit where water is likely to collect; if necessary dig some drainage. The pit is lined with birch logs, planks or half logs. Fill the pit with fresh cow manure from barn or pasture, free of straw. The cows should be pasture-raised or have been fed hay.

The biodynamic preparations are introduced in the usual manner, in holes about 20–30 cm (8–12 in) apart. Use a pinch (2 g) of each preparation and sprinkle with valerian (507), prepared as described earlier (2–5 ml, ½–1 tsp, of valerian stirred for a few minutes in 500 ml (17 fl oz) of warm water). Cover the pit with boards, branches or straw to avoid it from drying out or having too much contact with the rain. Re-introduce preparations roughly every 2 to 4 weeks.

Application and storage

When the content of the pit has become darker in colour, is malleable and crumbly and has lost most of its odour, it can be used in very small quantities like a starter that is mixed into manure, vegetable compost, animal bedding or liquid manure. The vegetable compost pile can be watered several times a week with a blend made up of one part birch pit compost to seven parts warm rain water and stirred for about 10 minutes. Birch pit compost is particularly useful when there is insufficient manure for a market garden.

Storage should be as for Maria Thun's barrel preparation (see p. 78).

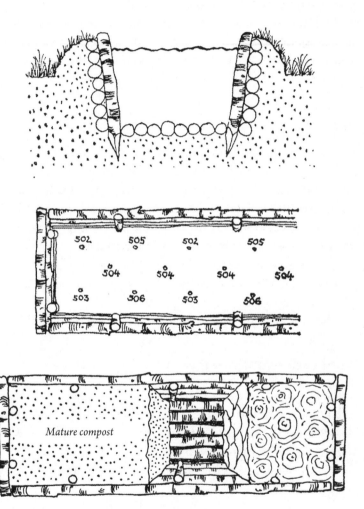

Birch pit preparation after Christian von Wistinghausen

Compost manure with nettle

Pioneered in the USA by Walter Goldstein, composted manure with nettle seems to give better yield results compared to using other composted manure. Its regulating effect on growth appears to be excellent, as well as its effect on the health of the plant. It particularly helps prevent peach leaf curl.

Method

Harvest green nettles ready to flower or just beginning to flower (0.5–2% of the weight of the manure). Chop them into small pieces, about 2 cm (1 in) long. With a shovel, thoroughly mix the manure with the nettle for 5 to 10 minutes. Put this mix in a pit and add the compost preparations 502 to 507, as for the barrel preparation. It will be used at 0.2 g of preparation per kg of manure.

If, instead, you use 0.4 g of preparation for each kg of manure (1/8 oz for each 20 lb), reduce the quantity of fresh nettle to 0.5% of the total weight.

The pit can be made like the birch pit or in a wooden barrel.

Application and storage

Once this compost has achieved the right crumbly consistency, like humus, it can be used like the other composted manures at a dosage at 240 g/ha, stirring for 20 minutes in 35–50 l of warm rainwater (3½ oz/acre in 4–5 gal).

Storage should be as for Maria Thun's barrel preparation (see p. 78).

Starters for kitchen compost

For environmental reasons, and the need to accumulate carbon in the soil to limit the negative effects of carbon dioxide in the atmosphere, we need to treat organic waste with care. Garden waste and cuttings and kitchen scraps need to be transformed into humus.

A compost starter helps domestic composting and favours the formation of good humus. Making the starter is a complex process requiring the best quality ingredients, including cow manure, clay, rock powder and the six biodynamic compost preparations. The best option, then is to purchase it from you local biodynamic association, where possible.

Method

The starter should be a dry, grey, fairly coarse powder. Put 10–12 g (1 tsp) of compost starter into a bucket containing half a litre of warm water (about 35°C, 95°F). Add a shovel of old compost or very good garden soil. With a trowel or stick mix it all together, alternately in one direction and then the other for several minutes.

Spread the mix evenly on the pile of accumulated organic waste. The waste should be aired but slightly damp. If it is too wet, add some old compost or fairly dry potting soil. Shake the starter like pepper over the compost pile.

A wooden bin for kitchen waste to be composted

As soon as that is completed, cover the pile with a thin layer of soil, or even straw or old hay.

Check the temperature and humidity regularly over the next couple of weeks, turning it over if needed, especially if it is too damp.

You can repeat the whole operation at regular intervals (every two to four weeks) depending on the quality of compost scraps that are added to the pile.

After several weeks or several months, depending on the season and the materials used, the compost will have a black colour and a crumbly consistency. It should have a good smell like the forest floor, and be full of compost worms. It is then ready to use.

Another way of using the starter is to keep some in your kitchen, adjacent to your kitchen caddy (if you use one, to collect vegetable and other kitchen waste before emptying in onto your compost pile). When the caddy is empty, sprinkle some old dry

Wooden bin for kitchen waste

compost in the bottom, and a pinch of starter. The compost will help soak up liquid from the kitchen scraps, preventing rotting which can cause unpleasant odours. The starter will help with maturation of the organic material into humus.

One advantage of this substance is that it is easy to store (when dry) for several months, in contrast to the biodynamic preparations or the classic composted manures that require great care and special storage boxes lined with peat.

The starter doesn't replace the six biodynamic compost preparations, but this method is simple, quick and affordable, accelerating the breakdown of household compost without using the preparations. It is a good introduction to biodynamic composting.

Animal bedding in a barn

5. Liquid Manure and Animal Bedding

Liquid manure in a slurry pit and animal bedding should be treated with biodynamic compost preparations. Regular use of these preparations significantly improves both liquid and solid manure, complementing classic techniques like aeration and introduction of clay minerals. The liquid manure will become more oily, more colloidal with less smell, and will be less toxic for the soil and plants.

First method

Use composted manure with added bentonite. The mix is equal parts of 240 g of compost preparation and 240 g of bentonite for 10–15 m³ (8½ oz for 350–500 cu ft) of animal bedding or liquid manure and stirred for several minutes in (preferably lukewarm) water, before emptying it into the pit. Do this at regular intervals, weekly or monthly depending on how fast you want the manure to mature.

Second method

Make a cross with two pieces of wood fastened at the centre and suspend a set of the five biodynamic compost preparations (502-506), a pinch (2 g) of each for 10–15 m³ (8½ oz for 350–500 cu ft). Each preparation should be wrapped in a small cotton or flannel cloth. The cross is then lowered into the pit so that the cloth bags soak in the liquid. To finish, stir a dose (2–5 ml, ½–1 tsp) of valerian for 10 to 20 minutes in 4–5 litres (1–1½ gal) of warm water, and pour this into the pit. This technique has been proven effective on several farms.

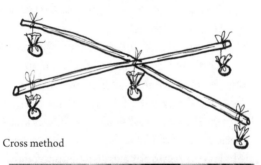

Cross method

Stirring the valerian preparation

Use in the barn and on accumulated animal bedding

The compost preparations can be used on animal bedding during winter. That allows the prepared manure to be spread in the spring. This saves a lot of time and energy. It also seems to reduce the loss of important substances such as nitrogen and other nutrients. Having said that, it is not to be used on crops prone to fungal disease or sensitivity to nitrogen: for example, pastures, heavy feeding crops like beets or corn, or green cover crops.

Measure the area of the barn and the depth of the bedding, and calculate the volume of accumulated manure. Insert a set of preparations for every 10 cubic metres (350 cu ft). Split the area into sections equivalent to a volume of about 10 m³, with a surface area around 20 to 40 m², and with a bar make six holes around each section. A pinch of each preparation is inserted without the need to wrap it in old compost as the temperature of the pile is never very high. Mix valerian in water and stir for several minutes. Add a small quantity into the sixth hole, and sprinkle the rest over the entire surface with a brush or a sprayer. This whole process can be done in the calm presence of the animals, and can be repeated several times over the course of the winter.

This practice has the merit of both rotting the manure into humus and suppressing ammonia odours. The drawback is that the bedding breaks down faster, so more straw than usual is required. The barrel preparations can also be sprayed over bedding to support maturing, as described in the previous chapter.

Manure for pasture, market garden and general gardening

Fill a container (like an old bathtub, or a vat that has not been used for chemicals) a quarter full of liquid manure or a dilution of solid manure. The more diverse the manure is, the better. Fill the rest up with water. Make a cross bar with two pieces of wood,

Galvanised plate mixer, 15–20 cm (6–8 in) in diameter, with 2 cm (¾ –1 in) holes and a long handle

and hang a set of the five biodynamic compost preparations, wrapped in cloth. Suspend the cross over the bathtub so that the preparations are just immersed (to about a third). Stir a portion of the valerian in 4–5 litres (1–1½ gal) of warm water for 10 to 20 minutes and pour the liquid in the container.

Stir the liquid twice a week. You can use a cream mixer or paint mixer with a shovel handle attached and stir it vigorously up and down to blend it.

After 2 to 4 months (depending on the time of year) the liquid turns black and oily. That is time to use it. Dilute it with 3 or 4 parts water before using it. Judging the right time and dilution is a matter of experience.

This liquid does not damage plants like a liquid fertiliser would since the substance has become colloidal. However it should not be used excessively. It is mainly for market gardens and crops where compost is rarely used. It can also be used on pastures with heavy clay and lime soils.

Whitewashed young apple trees

6. Pastes and Root Dip

In the Agriculture Course, Rudolf Steiner referred to human beings as upside-down plants, with a correlation between plant roots and the human head (the sensory and nervous system).

Tree trunks and vine stocks are therefore a bit like soil which has been raised up and given life, and as such they should be cared for in the same way as soil. They need clay, organic material (fresh manure, prepared compost, birth pit preparation), regulating elements (horsetail, whey), and other substances depending on their needs (basalt, maerl, sulphur, pigeon droppings, propolis, sodium silicate (waterglass), etc.).

The tree pastes and sprays described below are used for fruit trees, vines, berry bushes, roses, hedges, etc.

For trees, using paste on the trunk and large branches, and spraying all the branches after the leaves have fallen off, before bud burst or after winter pruning are essential and effective long-term measures for a well balanced, healthy orchard.

It is the same for vines, which should be sprayed after the leaves fall off, covering the vine stock. A mix of clay, fresh manure tea, horsetail decoction and sodium silicate can be used, optionally adding propolis or essential oils. After pruning, apply a consistent coating of a mix of clay and manure (with other optional ingredients) to large pruning wounds, with a paintbrush or spatula. Alternatively, spray the autumn mix using a backpack sprayer at the end of each pruning day. In difficult circumstances, the same mix can be sprayed on the vines at bud burst; in this case, add a little sulphur.

There are three mixes for different uses. The main difference is in the consistency.

- ⚕ A **thick paste for the trunk or large branches** can be applied with a paintbrush, a regular brush or a garden hose with nozzle. For amateurs with small orchards, it is a simple measure to apply to all trees. On a professional scale, the measure can be applied to at-risk trees or sections of the orchard, and to yearlings.
- ⚕ A **distemper** for trunks, large branches and crowns of trees as well as for berry shrubs, raspberry bushes, vines, etc. This spray is more liquid and is easily applied with a sprayer with a strong jet stream.
- ⚕ A **thick protective paste** helps heal pruning and grafting scars. It is similar to the first; only its use differs. It can be applied with a paintbrush or spatula.

Composition

The ingredients and their respective proportions vary a lot according to the recipe but the base is always the same: cow manure and clay. The cow dung should be without straw and preferably from pasture-fed (or organic/biodynamic fodder-fed) cows. Potter's clay (kaolin or ordinary clay) will bind better and is more durable than montmorillonite or bentonite (green or brown clay) which are quickly washed away by rain. Combine several clays together to capture their complementary properties. Mix clay in water to get a thin batter consistency. Sodium silicate at a dose of 1–2% can be added: it is good for protecting bark, and helps the paste to stick. It should be added to the mix at the end, and it is caustic so take care with glasses and tractor windows, and check the warnings on p. 166.

To make a fluid paste of the required consistency, add one or more of following liquids:

- ⚘ ideally, horsetail decoction;
- ⚘ otherwise, nettle tea or other fermented herbal teas;
- ⚘ whey or skimmed milk;
- ⚘ rain water;
- ⚘ some biodynamists use horn manure (500 or 500P) or barrel preparation. If you are using any of these, they should be stirred for one hour or 20 minutes before (depending on the preparation) and used within 2 hours of stirring (barrel preparation can be kept for up to 72 hours).

If necessary, add different minerals: ash, maerl, basalt, potassium salts, diatomaceous earth (*kieselguhr*), potassium permanganate against moss and lichens, as well as various things like pigeon droppings or propolis.

Ideally apply the tree paste or spray the entire tree in November after the leaf-fall and again in February or March before bud burst. If these applications are carefully done, winter treatments and white oils will be unnecessary.

In difficult situations with a heavy contamination of mushroom spores, pests or scale insects, first spray lime sulphur at 20% dilution at 500–1000 l/ha (55–110 gal/acre).

Recipes for pastes and distempers

Australian method

Put fresh cow manure in a thick canvas bag. Put the bag in a large vessel made of copper, stainless steel or wood and fill with pure water to soak. Possibly add a little pigeon droppings. Let it ferment for 24 to 48 hours, and filter before use.

In another vessel, mix kaolin in enough water to make it swell and form a sort of soup. Kaolin is a good binder for coating (green clay, bentonites and montmorillonites are easily washed away by rain, so it's better to use them mixed with kaolin).

Add the filtered cow manure to dilute. You can also add sodium silica (maximum 2%) and wettable sulphur. Add water to make a total volume of 250–400 l/ha (25–40 gal/acre). Cover the entire tree with this preparation before winter and after pruning, and in any case before bud burst. Spray with a backpack sprayer, or with a tractor sprayer.

Ehrenfried Pfeiffer's method

Mix 2 buckets of cow dung, 2 buckets of clay, 1 bucket of *kieselguhr* (diatomaceous earth). Decoct 250 g (9 oz) of horsetail in 10 l (2½ gal) of water, and add to the mix. Add 200 g (7 oz) horn manure (500) stirred in 100 l (26 gal) of water. After sieving, use on crowns, branches and trunks on about 2 ha (5 acres) of orchard.

Volkmar Lust's method

Add 5 kg of kaolin, 3 kg potassium sulphate, 700 g of wettable sulphur and 1 l of sodium silicate to 100 l of water (11 lb, 6½ lb, 24 oz, 1 qt to 26 gal). This is a good mix for general use; if being used as a preventative treatment for scabbing, this should be done in November and in the spring before bud burst. About 800–1000 l/ha (85–105 gal/acre) are needed.

Pastes for wounds and care for the trunk

Make a mix of kaolin and cow dung in equal parts into a paste. Add 1% horsetail tea decoction that has been stirred for 20 minutes. The quantity depends on the quality of the coating and the number of trees. If horsetail is not available, use 2% sodium silicate.

Apply this paste where the bark is loose or wherever a tree is in poor health. It can be applied on tree wounds or young trees that have had difficult growth, after removing the dead parts of the bark. It is also a good paste for pruning scars.

Paste on a large pruning scar

The paste is applied with a paintbrush. It can be stored for several days, but when it starts to smell off, it is time to stop using it.

Maria Thun's paste

Mix equal parts of cow dung, clay and whey. Once the mixture is sufficiently fluid, coat the cleaned-off trunk with a brush, preferably in November. It can be reapplied in February or March.

Volkmar Lust's paste

For about 50 low-branch trees about 10 years old, with crowns in goblet shape and trunks roughly 60–80 cm (2–3 ft) high, use:

- ↯ 5 kg (11 lb) kaolin to stimulate and rejuvenate the trunk and protect from frost damage (the clay acts as a mediator between cosmic and earthly forces),
- ↯ 3 kg (6½ lb) cow manure, to stimulate the life of the trunk by different fermentations and hormones,
- ↯ 500 g (1 lb) maerl which contains calcium, magnesium and trace elements of marine origin,
- ↯ 500 g (1 lb) micronised basalt which contains calcium, magnesium, silicic acid and trace elements of volcanic origin,
- ↯ 500 g (1 lb) potassium sulphate that acts on moss, lichens and animal pests.

Mix in a tub with 12 l (3 gal) of lukewarm rainwater into a homogenous blend. Then add 500 g (1 lb) sodium silicate which thickens the paste and increases the binding effect. It also acts on moss, lichens and fungi.

Paste with pure clay or diluted with whey

Use this paste as a poultice for large wounds on the bark and trunk cavity.

Sprays for vines

Frédéric Lafarge's spray

Use after the leaves have fallen off. Spray 250 l/ha (25 gal/acre). These quantities are for 12 hectares (30 acres):

- ❦ Manure tea made with 50 kg (110 lb) good quality fresh cow-manure. You can add a bucket of chicken droppings if you have some;
- ❦ Horsetail decoction made with 1.2 kg (2½ lb) dried horsetail (Equisetum arvense), gently heated in 50 l (13 gal) of rain water for 45 minutes. Macerating the plant the day before reinforces its activity;
- ❦ 210 kg (450 lb) bentonite clay (Lafaure) diluted in rain water to be about 7% of the total spray;
- ❦ 60 kg (130 lb) 2% sodium silicate as a bonding agent, added at the last moment (use with care).

Spraying with an atomiser (Frédéric Lafarge)

François Duvivier's spray

Version 1: use in the autumn after the leaves have fallen off. Spray 300 l/ha (30 gal/acre).

- 15 kg/ha (13 lb/acre) of clay diluted to 5% concentration, 24 to 48 hours beforehand;
- Horsetail decoction made from 100 g (3½ oz) dried horsetail in several litres of rain water;
- 4 kg/ha (3½ lb/acre) good quality fresh cow-manure tea macerated for 24 to 48 hours;
- 25–30 l/ha (3 gal/acre) whey;
- 50 ml/ha (3½ tsp/acre) propolis;
- 6 l/ha (2½ qt/acre) 2% sodium silicate added at the last moment.

Version 2: use after pruning and before bud burst. Spray 150 l/ha (15 gal/acre):

- Same as version 1. Less spray is used but increase the clay concentration to 7%.

François Bouchet's spray

This is a light spray based on Maria Thun's barrel preparation, whey and bentonite clay. It is used in autumn after the leaves have fallen off, and again after pruning.

For one hectare, use 240 g (8½ oz) barrel preparation, 2 l (2 quarts) of whey and a light dilution of clay, about 1–2%. Stir for 20 minutes, filter, then spray at 120–200 l/ha (12–20 gal/acre), within 48 hours of preparation. Spraying equipment used for copper and sulphur sprays is suitable.

This spray is easy to use but less effective than heavier sprays with more clay, which affect the soil as well as the plant.

However, compost tea made from fresh cow manure has a particularly beneficial effect on wood and bark.

Frédéric Lafarge's spray for vine pruning scars

Spray 90 l/ha (9 gal/acre). A 15 l (4 gal) backpack sprayer will treat about a fifth of a hectare (half an acre).

Make a horsetail decoction two weeks before use, decocting 300 g in 10 l water (10 oz in 2½ gal) for 45 minutes. Store in a cool, dark place. Make the clay mixture a week before, using 4.2 kg bentonite (Lafaure) clay in 60 l water (9 lb/16 gal).

Mix just before spraying. Blend 12.5 l (3½ gallons) clay diluted to 7% concentration, 1 l (1 qt) of the prepared horsetail decoction, 10 ml (3½ fl oz) of propolis tincture and, at the last minute, 300 ml (10 fl oz) sodium silicate.

Spray at the end of each pruning day or half day using a backpack sprayer.

Marc and Pierrette Guillemot's paste for vine pruning scars

At the end of each half day of pruning, take half an hour to coat the scars, applying following mix with a paintbrush.

Mix half fresh cow manure and half potter's clay, then add whey or, preferably, horsetail decoction to the consistency of very thick paint. Sifted wood ashes or micronised basalt can also be added. The mix can be prepared two or three days in advance (but no more than that).

This paste can't be used when the sap is flowing strongly. The paste can also be further diluted and sprayed; whey used to be used for dilution, but horsetail decoction seems better.

Spray for large pruning scars

Mix several drops of rosemary essential oil in one litre (one quart) of water, with a little whole milk. Spray onto scars with a small hand sprayer. Scars will dry out within 24 hours.

Root dips

Root dip is made with a mix of one quarter fresh cow dung and three quarters clay, or clay soil from the site of transplanting. Add horn manure (500) or prepared horn manure (500P) and stir for one hour. Add sufficient water to make a paste with the consistency of a thick pancake batter which will stick well to the roots.

Root dip is useful for all forms of transplanting, in market gardens as well as orchards. It is important to put the roots of a new plant into a root dip as soon as you receive the plant, especially if you can't transplant the plant immediately (for example, if the time is unfavourable according to the planting calendar).

Depending on the situation, in non-calcareous soil, add some maerl to the mix, or in some cases, basalt powder (as used in gardens).

Other practices use an equal quantity of manure and clay, even adding a decoction of horsetail. If you cannot get fresh cow dung, use good biodynamic compost.

Root dip for vines

Before transplanting vines, spray prepared horn manure (500P) on the planting holes and surrounding soil. This can be done a day or two before transplanting. The 500P should be prepared with 35–100 g (1–3 oz) prepared horn manure (500P) stirred in 35 l (9 gal) of rain water.

For the root dip, mix three buckets of cow dung and three

buckets of bentonite clay with a little water. In late afternoon, stir 135–200 g (5–7 oz) prepared horn manure (500P) in 75 l (20 gal) of rain water. Mix immediately with the clay mixture until you have the consistency of thick pancake batter. The dip should be used as soon as possible (at most, 2 hours after dynamisation).

There are different ways to use the dip:

- take the plants out of their bags and soak overnight in the root dip; or
- pour root dip directly into the bag, ensuring it covers all the roots (1 to 2 l (1 to 2 quarts) per bag). The plants can remain this way for several days before planting.

7. Working with Cosmic Rhythms

Working with the rhythms of the earth and the cosmos is part of biodynamics. An increasing number of studies demonstrate the influence of the stars on the living world. A lunar and planetary planting calendar (for instance, *The Maria Thun Biodynamic Calendar*) is therefore very useful for observing the effect of rhythms on plants and are a useful tool for planning work. However, a calendar should not be followed too strictly or rigidly, especially when applying the preparations. It should not be used to treat the plant in an unswerving or unquestioning manner; for example, only spraying vines on days which are favourable for fruiting (fruit days in the calendar). All plants need to be allowed to develop in sequence, through the phases of root, leaf, flower and fruit.

The basic rhythms of the moon and planets are explained in the calendar. Here we shall assume some knowledge of these astronomical rhythms. When sowing, working with the soil, composting or using the preparations, **avoid working on days where there are nodes** (☊ or ☋ in the calendar). The days preceding and the hours around the moon's perigee (pg in the calendar), especially if close to a full moon, require particular care for plant health and protection in the garden, orchard or vineyard. For plants that are prone to fungal disease, these times should be used for preventative treatments (horn silica, horsetail decoctions or various teas). In most cases, however, perigee is an excellent time for sowing and for work that stimulates vegetative (growth) activity and the multiplication of cells (see in particular the work of Hartmut Spiess in Darnstadt).

For sowing and transplanting, follow the descending moon rhythm (when the moon passes from the constellation of Gemini to Sagittarius in the northern hemisphere). For composting, working during the descending moon time is more important in the spring than in autumn. For winter pruning, the descending moon is favourable to give back vitality to vines and weak trees; winter pruning during the ascending moon is good if vines have becomes too vigorous – especially because rhythms repeat over several years.

The rhythm of the day (breathing out in the morning and breathing in during the evening) is really important and can help crops develop better. Working the soil (harrowing) in the morning can release excess humidity, while working the soil in the evening conserves water during dry periods. The same goes for harvesting: taste qualities and storage are improved by harvesting above-ground plants in the morning, preferably during the ascending moon, and harvesting root vegetables in the evening, preferably during a descending moon.

When using the preparations and doing critical work such

March 2014

All times in GMT

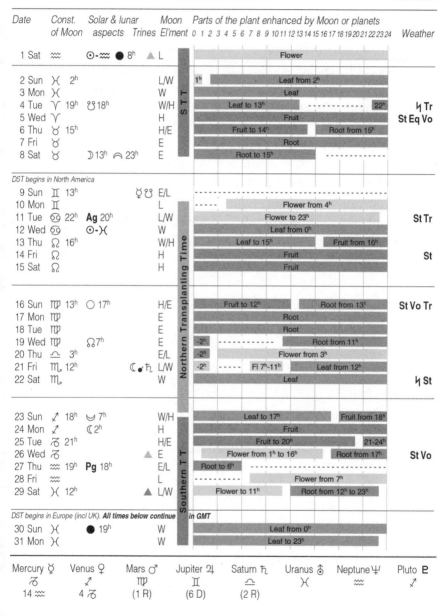

Date	Const. of Moon	Solar & lunar aspects	Trines	Moon El'ment	Transplanting time	Parts of the plant enhanced by Moon or planets (0–24)	Weather
1 Sat	♒	☉-♒ ● 8^h	▲	L	S T T	Flower	
2 Sun	♓ 2^h			L/W	S T T	1^h; Leaf from 2^h	
3 Mon	♓			W	S T T	Leaf	
4 Tue	♈ 19^h	☋ 18^h		W/H	S T T	Leaf to 13^h - - - - - 22^h	♄ Tr
5 Wed	♈			H	S T T	Fruit	St Eq Vo
6 Thu	♉ 15^h			H/E	S T T	Fruit to 14^h; Root from 15^h	
7 Fri	♉			E	S T T	Root	
8 Sat	♉	☽ 13^h ⌒ 23^h		E	S T T	Root to 15^h - - - - -	
DST begins in North America							
9 Sun	♊ 13^h		☿☋	E/L		- - - - -	
10 Mon	♊			L	Northern Transplanting Time	- - - - - Flower from 4^h	
11 Tue	♋ 22^h	**Ag** 20^h		L/W	Northern Transplanting Time	Flower to 23^h	St Tr
12 Wed	♋	☉-♓		W	Northern Transplanting Time	Leaf from 0^h	
13 Thu	♌ 16^h			W/H	Northern Transplanting Time	Leaf to 15^h; Fruit from 16^h	
14 Fri	♌			H	Northern Transplanting Time	Fruit	St
15 Sat	♌			H	Northern Transplanting Time	Fruit	
16 Sun	♍ 13^h	○ 17^h		H/E	Northern Transplanting Time	Fruit to 12^h; Root from 13^h	St Vo Tr
17 Mon	♍			E	Northern Transplanting Time	Root	
18 Tue	♍			E	Northern Transplanting Time	Root	
19 Wed	♍	☊ 7^h		E	Northern Transplanting Time	-2^h - - - - - Root from 11^h	
20 Thu	♎ 3^h			E/L	Northern Transplanting Time	-2^h; Flower from 3^h	
21 Fri	♏ 12^h		☾☌♄	L/W	Northern Transplanting Time	-2^h - - - - - Fl 7^h-11^h; Leaf from 12^h	
22 Sat	♏			W	Northern Transplanting Time	Leaf	♄ St
23 Sun	♐ 18^h	◡ 7^h		W/H	Southern T T	Leaf to 17^h; Fruit from 18^h	
24 Mon	♐	☾ 2^h		H	Southern T T	Fruit	
25 Tue	♑ 21^h			H/E	Southern T T	Fruit to 20^h; 21-24^h	
26 Wed	♑		▲	E	Southern T T	Flower from 1^h to 16^h; Root from 17^h	St Vo
27 Thu	♒ 19^h	**Pg** 18^h		E/L	Southern T T	Root to 6^h - - - - -	
28 Fri	♒			L	Southern T T	- - - - - Flower from 7^h	
29 Sat	♓ 12^h		▲	L/W	Southern T T	Flower to 11^h; Root from 12^h to 23^h	
DST begins in Europe (incl UK). **All times below continue in GMT**							
30 Sun	♓	● 19^h		W	Southern T T	Leaf from 0^h	
31 Mon	♓			W	Southern T T	Leaf to 23^h	

Mercury ☿	Venus ♀	Mars ♂	Jupiter ♃	Saturn ♄	Uranus ♅	Neptune ♆	Pluto ♇
♑	♐	♍	♊	♎	♓	♒	♐
14 ♒	4 ♑	(1 R)	(6 D)	(2 R)			

as harvesting, sowing, etc., avoid the period from local noon to 3 pm (the three hours following the high point of the sun). Stirring and spraying of preparations aimed at soil and roots should only be done in the evening.

For vineyards and orchards, as well for seed production, horn silica can be applied three times, at intervals of eight days, when the moon is in a fruit constellation. **However, it is always more important to observe the condition of the soil, and weather conditions take precedence over times recommended in the planting calendar.**

Always note which constellation of the zodiac is prevalent when working, spraying, etc., and carefully note the effects on the soil and the plants. These observations and notes allow you to take stock at the end of the season and can be used in subsequent years.

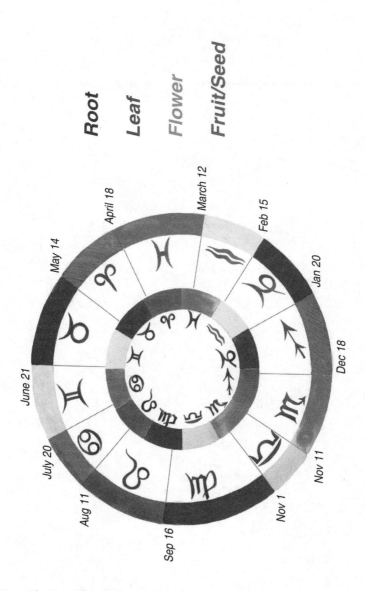

From *The Maria Thun Biodynamic Calendar*

~ Part 2 ~

Supplementary Practices

8. Plant Extracts, Herbal Teas and Decoctions

The ideas in this chapter come from the agricultural experience of many biodynamic practitioners. In most cases, these methods only give good results where there is already good agronomic, organic and biodynamic practice. The correct environmental conditions must first be established on the farm or the garden, which include caring for the land, using appropriate seeds and manure, and creating the greatest possible diversity of animals and plants. Biodynamic preparations, both for spraying and introdced into compost, are a basic requirement for favourable plant health; it is this basis which makes methods using plant extracts, teas and decoctions effective.

Plant extracts can, however, help prevent a number of pests and crop diseases, and in some cases are curative. Many wild or cultivated plants have significant potential to stimulate growth as well as to regulate or heal various crop disorders. These healing plants can often be found locally and are frequently adapted to local conditions, which helps prevent diseases or imbalances specific to the locality.

The quality of water used to make extracts, teas and decoctions is extremely important. The best is water that is slightly acidic (pH 6 to 6.5) and that has a low mineral content, like rainwater. While for the biodynamic preparations it is not desirable to adjust the pH of the water, for teas and decoctions it may need adjusting. Test with a pH metre or less accurately with litmus paper. Water from a spring or river in limestone regions can be acidified by adding anywhere from 2 glasses to several litres of cider or wine vinegar to 100 l of water. In granite-rich regions where the water may be too acidic, add some maerl to achieve the right level. (Adjusting the acidity of water is sometimes also necessary for other biodynamic treatments such as Bacillus thuringiensis, copper or sulphur treatments, organic insecticides, etc; see chapter *Products for Stimulating and Regulating Plant Health.*)

These healing herbs and plants can be cultivated and harvested when needed. Choose plants of organic or, preferably, biodynamic origin. If wild harvesting, find a unpolluted site. Carefully observe the quality of the crop; the active properties of plants are different depending on the time of year. Most plants are harvested at the beginning of the flowering stage, preferably in the morning of a nice sunny day, after the dew has lifted.

However, there are some exceptions. Stinging nettle is best picked just before sunrise, and in the spring. Dandelion should be picked in the late morning, early in spring, in good weather, before the centre of the flower has opened up. Pick valerian inflorescences in the morning or the later afternoon (not early afternoon), keeping them dry and free of rain or dew. St John's wort is usually picked while the sun is at its strongest, at midsummer in the middle of the day. Horsetail is best picked from the end of June to August. Willow is ideally collected early in spring when salicylic acid levels are highest. Oak bark should be harvested in the autumn when the calcium content is greatest. Bark should always be collected in the afternoon.

Taking account of lunar and planetary influences (from the planting calendar) can improve drying, storage and efficacy.

Speedy, but not excessive, drying of plants and herbs is important. It should always be done in the shade, such as in a well ventilated barn, or over a stove or even a dryer, but not exceeding 30–35°C (85–95°F).

Gas-heated tea maker

Wood-heated tea maker

Gas-heated tea maker

113

Good storage is essential. As soon as the plants are well dried, put them in cloth or paper bags, or into metal or even cardboard boxes that stay dry. Dried plants do not keep more than a year to a year and a half at most.

Before use, the plants must be 'opened' to access their active properties.

There are different methods of preparation:

Cold ferment

Cold ferment (or slurry) is soaking a solid in water to soften it or break it up. Depending on the required properties, it can be soaked for shorter or longer periods. It is used for stinging nettle, comfrey, fern, etc.

Tea (infusion)

Tea (or infusion) is made by putting herbs in cold water and heating. As soon as the water boils, remove from the heat

Nettle ferment

source and leave the plants for 10 to 20 minutes before using. This method is generally used for plants whose active part is the flower or leaf. Some plants, like willow and queen of the meadow *(Spirea ulmaria)* must not be brought to the boil, as their active properties would be destroyed.

Decoction

Decoction is a longer cooking method (from a few minutes to an hour), over a low heat and covered. It is used for plants whose active properties are more difficult to extract, such as horsetail, certain barks and roots.

All these plant-based preparations can be stirred (dynamised) for 20 minutes; this will gently increase their efficacy especially if they are being used on their own. Stirring is unnecessary if the plant extracts are to be mixed with products like sulphur or copper. Most of the plant extracts can be used in a mix, but their preparation should generally be done separately, plant by plant. There are some exceptions mentioned below.

Primary plant extracts

Horsetail decoction *(Equisetum arvense)*(508)

The properties of horsetail decoction as an anti-fungal agent were mentioned by Rudolf Steiner in the Agriculture Course in 1924. It is used as a preventative measure against mildew, rust, brown fruit rot and other fungal diseases. In a number of cases, it has been shown to have a powerful curative property.

The characteristics of horsetail can help us understand its effects. It is a master of damp conditions and hence is good at preventing fungal and watery lunar influences from attacking the aerial parts of plants. This is also useful knowledge for limiting

Horsetail: on the left, a fertile stalk; on the right, a sterile stalk

the spread of spores. Horsetail got rid of the desire to flourish (grow flowers) and the instinct for reproduction (develop seeds), and lives in a serene, purely vegetative state; similarly this is the outcome farmers hope to see against fungus, especially mildew.

Horsetail is rich in silica which improves the quality of plant tissues, but take care with its use during the growing season since it can dry out soil and plants. In this case, use in conjunction with willow and nettle.

Horsetail decoction can be used on grain crops, in the garden, in orchards and vineyards. According to Harald Kabisch, it should be sprayed in the morning, and it must be used two days before full moon for optimal activity. As a preventative, spray the soil in spring from March to May and in autumn in October or November. This holds true for gardens as well, especially for potatoes. In viticulture it is a proven preventative used the week before Easter. It can also be used after pruning and painting vines as well as trees.

On vines and fruit trees, it is valued as a general preventative against fungal disease when sprayed at the bud burst. Another application in the autumn, when the leaves begin to fall, combats wood diseases. It will save time to add it to the distemper when painting trees at the beginning and end of winter.

It can be used frequently in times of heat and humidity in spring, especially close to full moon and the moon's perigee, times of increased fungal disease. But during the summer, take care in viticulture and fruit trees when using the horsetail decoction, as it can have quite a drying effect, especially on sandy soil and soil rich in silica.

In the market garden, good results have been shown in preventing bean rust, potato and tomato blight, bacterial disease and lettuce fungus.

On winter grain crops, it is useful at the beginning of the autumn season and on barley as the start of tillage. Its use should be repeated several times in succession (as noted by Peter Kunz, a seed selector in Switzerland).

When applied on hail damage, it has sometimes has positive effects, but it is preferable to use nettle, or better, valerian.

Preparation

Sterile stems of horsetail should be harvested at maturity, between the end of June and the end of July, or even at the beginning of August when it has reached its maximum silica content. Make sure that you are harvesting the right species, *Equisetum arvense*, easily identifiable by its tapered stems (marsh horsetail, *Equisetium palustre*, and great horsetail, *Equisetum telmateia*, do not have the same beneficial properties). The sterile stems are 20–80 cm (8–30 in) long, and rough to the touch due to the silicate content. The top part of the outer girdle is divided in 8 to 12 whorls. The branches do not droop, are full and are four-sided in section.

The plants are effective only for a year after have been harvested, so it is pointless to stock up on them. The upper sterile stems are used either fresh or dried.

With dried plants use 100–120 g/ha (1½ oz/acre). Cook this over a medium heat in 3–5 l (¾–1¼ gal) of rain water, allowing it to come to a light boil for 40 to 60 minutes in a covered container. Macerating the plant the day before reinforces its activity.

With fresh plants take about 800–1000 g (1¾ – 2¼ lb) of fresh leafy stems of horsetail and cover with rain water. Heat until a few bubbles appear then simmer gently, covered, for 40 minutes. To use this decoction, let it cool and filter until you have a solution resembling tea. Dilute 1 part decoction to 9 parts water.

Stir in the same way as 500 or 501, but only for 20 minutes, just before spraying to increase its efficacy. The undiluted decoction can be stored for several weeks in a cool dark place, in a tinted glass container.

Use 35–70 l/ha (4–7½ gal/acre), depending on the spraying equipment available. If used during the vegetative phase, it should be directed under the foliage with fine nozzles, with sufficient pressure.

Warning

If this preparation is used too intensely (more than 10 passes on the same crop) it is inclined to disrupt the soil which needs to build up its own fungal flora (actinomycetes, etc.)

Some trials using horsetail decoction at a dilution of D5 have shown good results against mildew. Take care, though, since it can damage weak plants or weak soil. A lot of research still needs to be done to find optimal dosage and dilutions of this decoction.

Horsetail *(Equisetum arvense)* is a plant that can become a weed. Its introduction in certain regions, where it is not indigenous, is not encouraged. (In New Zealand where it was introduced in the 1920s, it has been declared an invasive species preventing its sale or cultivation.) If it has to be cultivated, it should be well controlled, for example, grown in old concrete drains.

Substituting horsetail with casuarina

Casuarina equisetifolia is a tree native to south east Asia, northern Australia and Madagascar. It is commonly known as horsetail tree, Australian pine, filao tree and other names. It can be used as an alternative to horsetail, though it is less effective. The male tree of the casuarina is used. Find a tree old enough to recognise its sex with certainty. Do not use a tree that has cones or one surrounded by cones on the ground, as this will most likely be a female tree. It is better to use a known local tree, since occasionally a male tree can develop small cones due to outside influences (such as a drought that induces it to reproduce).

Stinging nettle tea (*Urtica dioica*)

Stinging nettle tea is a regulator and stimulator of vegetative growth. It has a moderate preventative effect on mildew on vines and potatoes. On its own it generally does not prevent attacks of mildew: either incorporate it into a complex approach, together or alternating with other plant extracts and clay, or add small doses of copper. In viticulture, it controls spider mites. The tea is valuable in the spring to help the development of young shoots of grain.

Nettle at the right stage of flowering for tea

Preparing stinging nettle tea

Preparation

Use about four handfuls of fresh nettles (0.8 to 1 kg, (1¾ to 2¼ lb), or 100 g (3½ oz) of dry nettle for 3–5 l (quarts) of water. The best time to harvest is at the beginning of the flowering stage. Place the plants in cold water and heat. As soon as it boils, remove from the heat source and allow to infuse for 10 to 20 minutes. Then dilute it to make 35–50 l per ha (4–5 gal/acre). You can also add 0.5–1% (by weight) of clay (montmorillonite or kaolinite). According to Maria Thun, it is best sprayed in the evening on a leaf day.

This tea is used frequently in viticulture, and can be used in a mix with copper products and sulphur (the normal dose of copper or sulphur can then be reduced), although this is only possible if the pH of the mixture is neutral or slightly acidic. Do not add clay if mixing with copper, as they are incompatible.

Willow tea (*Salix*)

Introduced by François Bouchet, a biodynamic consultant, willow tea is effective in viticulture in preventing mildew. It is also effective against powdery mildew and botrytis. It acts like willow on its environment, overcoming humidity; the salicylic

Living willow

acid content regulates and mediates through the liquid sphere, particularly overcoming blockages of sap due to cold. It is best to use new bark from the beginning of the year as it contains more active ingredients. Harvesting is best in spring (February to April, before leaf formation).

We previously recommended 100 g of dried stems as the correct dose, but recent experiments have shown that a higher dose works better.

Preparation

Chop 100 g of dry bark or 200 g of dried stems (3½ oz, 7 oz) in a grinder, or manually with an axe or hatchet. Infuse for 15 to 20 minutes, in 4–5 l (quarts) of very hot water. Do not allow it to boil, as this would kill off the salicylic acid. The resulting liquid should be diluted to about 10% in 35–45 l (9–12 gal) of water in a barrel or in a tank sprayer, together with any additional ingredients (like copper or copper substitutes).

Use different kinds of willow for this preparation as diversity improves it *(Salix viminalis, Salix rubra, Salix purpurea, etc.).* The weeping willow, goat willow or other willows with large leaves are not suitable. Use willow in a mix with other plants, in particular with nettle. As for all teas and plant extracts described here, using a single or even several associated plants is not sufficient to prevent all fungal disease in vines and orchards. In practice, combining the extracts with small doses of copper and sulphur give positive results and a useful safety net.

Horsetail (*Equisetum arvense*) and nettle tea

Horsetail and nettle tea has given good results in the greenhouse and open fields as a stimulant for the natural defence of plants. It can be used regularly and up to twice a week in challenging conditions. In 2006 some trials were done on vines in Alsace,

France which showed that this mixed tea could limit the doses of copper needed to prevent of mildew.

Preparation

Put a good bunch of three quarters nettles and one quarter horsetail into 5 l (1½ gal) of cold water. Bring to the boil and simmer for 5 minutes, remove from the heat and infuse for another 10 minutes. Add 15 l (4 gal) of fresh water, filter, and fill a backpack sprayer; 18–20 l can cover up to one hectare (2 gal/acre). When using dried plants, use 50–100 g (2–3½ oz) of nettle mixed with 50–100 g (2–3½ oz) of horsetail. This tea can be used with copper and sulphur.

Valerian extract (507)

In the Agriculture Course, Rudolf Steiner said: 'press the blossoms of the valerian plant, *Valeriana officinalis,* and greatly dilute the extract with warm water. This will stimulate the manure to relate in the right way to the substance we call phosphorus.' Once known as a 'heal-all', it is a remedy given for stress, heart conditions and insomnia in humans. As a medicine, the rhizomes are used along with its roots, whereas in agriculture, only the flowers are used.

The preparation is obtained by extracting the juice of the flowers by pressing. This produces a dark extract with a strong animal smell. It can also be done by fermenting the petals in water in a glass jar and exposing to part-shade, part-light for 10 to 15 days. When the liquid begins to turn golden-green in colour and has a pleasant smell, it is time to filter, transfer to jars with lids, and store in a cool, dark place.

Harvesting valerian

Fermenting valerian

Use in agriculture

After stirring for 10 to 20 minutes in lukewarm water, the preparation is put into compost piles and sprayed over it as a cover (see earlier chapters for details). The preparation forms

124

a protective skin, and stimulates the development of compost worms.

Valerian is an extraordinary flowering stimulant. According to Maria Thun, use either the extract (507) or a tea made from dried flowers.

Legumes (peas, beans, field beans, clover, etc.) are particularly stimulated by a valerian treatment as it encourages the formation of nodes. According to Sattler and Wistinghausen, oilseeds (flax, mustard, canola, sunflower) as well as seeds of leafy plants (sainfoin, phacelia, etc.) can also benefit from its effects. Both yield and quality of seeds are improved if sprayed once when the plants are 15–20 cm (6–8 in) high, and then a second time before flowering. For the latter, adding some poultry droppings, particularly pigeon's, in the stirring vessel increases its efficacy even more.

Valerian is very useful in the spring, especially if the plants are stressed due to variation or change of climatic conditions. Add some to horn manure (500 or 500P) and to horn silica (501) at the beginning or end of stirring. According to Alex Podolinsky, it is one of the rare instances where we can mix biodynamic preparations for spraying. Horn silica stirred with valerian can be used in the spring in orchards (particularly on cherry trees) if there is an invasion of fungi after unusual cold spells. Use one drop per litre (quart) of water, or 5–10 ml/ha (2–4 cc/acre), stirred for 10–20 minutes, in the early morning. (To judge the right quantity, use the odour as an indicator; when stirring the liquid should smell slightly of valerian.)

In very dry conditions, irrigate the crop within an hour of spraying valerian, as spraying can cause a slight wilting.

If there is risk of frost (to –4°C, 25°F), spray it as a very fine mist in the evening beforehand on berry bushes, vines, fruit trees and sensitive vegetables (beans, tomatoes, basil, early potatoes. etc.). Use 2–10 ml/ha (1–4 cc/acre), stirred for 20 minutes in 30 l (8 gal) lukewarm water.

Valerian can also be used on vines and orchards that have difficulty fruiting. Choose a fruit day (according to the planting calendar) around St John's Tide (June 24), to encourage flowering and fruiting in the following year. Spray after any hail, together with a nettle tea or a couple of drops of arnica tincture, or use as a powder added to clay. It relieves the stress and shows its effects quickly. Spray as a fine mist, less fine than spraying horn silica (501), but aimed closer to the vegetation. The aim is to create an ambiance around the plants, misting the foliage very lightly, without really wetting the leaves. This is why quantities of less than 35 l/ha (4 gal/acre) are sufficient.

Recent research in Denmark while doing a grain study showed increase yields (weight per 1000 grains) following use of valerian stirred in water (10 ml in 40 l/ha, 1 tsp in 3 gal/acre) and sprayed four times during moon Saturn oppositions, or while the moon is in the constellation Cancer.

New vineyard research

Some useful research has been done by a group of winemakers in Burgundy on the use of valerian extract.

In spring, used with horn manure 500 or 500P, valerian can reduce the damaging effects of extreme daytime and nighttime temperatures, which are increasingly experienced.

The growth of young plants can sometimes be slowed down by the first spraying of horn silica (501) at the 5-leaf stage, especially if they're just budding. Adding valerian, before stirring the 501, at 5 ml/ha (½ fl oz/10 acres) reduces this effect.

Spraying valerian after trimming or pollarding reduces the stress to the vine and help prevent suckers and clusters. For this use, the valerian doesn't need to be stirred first.

Added to 501 and applied after flowering and before harvesting, valerian improves the quality of the grapes, encouraging better pip and phenol formation. An increase of phenol helps grapes to ripen more quickly, an important advantage in our changing climate.

It's worth noting that, contrary to much written advice, adding valerian to the preparations 20 minutes before the end of stirring actually produces negative results for vines, including poor leaf and branch growth and increased susceptibility to rot. Valerian should be added to 500, 500P and 501 before stirring, or at the very end.

Secondary plant extracts

Yarrow tea (*Achillea millefolium*)

Tea made from yarrow can reduce the amount of spraying done with sulphur in vineyards and orchards. According to Maria Thun this is also a very effective tea for grain, assisting health and promoting good reproduction for years to come. Spray at an early stage and again before harvest. Rudolf Steiner said this plant has a refreshing character, and can be beneficial during heat waves on all crops that are suffering. In summer, this tea is a good complement to other plant extracts such as nettle, horsetail and willow.

Yarrow tea used on vines during the growing season brings a 'sulphur signature' in the sensitive crystallisation pattern of the finished wine, with the wine less susceptible to oxidisation. The winemaker is therefore able to use less sulphur at other stages.

Preparation

Make the tea with 10 g of dried flower heads in 3.5 l (1 gal) of water. Dilute this to 35 l for one hectare (10 gal for 2½ acres). It is a good complement to other plant extracts.

Yarrow tea made of fresh leaves has a good result on Septoria leaf spot of the tomato plant.

Yarrow

Yarrow and nettle tea mix

Yarrow and nettle tea can do wonders in the vegetable garden controlling insects and fungi (suggested by Michel Leclaire, a market gardener in Troyes, France).

Preparation

Put a handful of fresh yarrow flowers in 10 l (2½ gal) of cold water and bring to a boil. Throw a little bucketful of fresh nettle or a couple of handfuls of dried nettle into the boiling water, then turn off the heat. Let it ferment and cool completely before using. It can be used pure or diluted.

Meadowsweet

Meadowsweet tea (*Spirea ulmaria*)

This is an excellent preventative for mildew, similar to willow tea. For one hectare (2½ acres), warm 250 g of dried flowers in 10 l cold water (9 oz in 2½ gallons). Don't heat beyond 80 °C (180 °F) or the salicylic acid will be destroyed. Copper and sulphur products can also be added.

Oak bark decoction (*Quercus robur*)

The oak bark decoction gently acts against most fungal diseases. It can be used to complement spring or autumn spraying of horsetail tea. According to Maria Thun, it particularly stimulates lettuce, radish, cauliflower and tomatoes. Do not use excessively, this can cause severe blockages in plant growth.

Only harvest the outer bark, preferably from a live or newly felled tree. Pedunculate oak (*Quercus robur*), sessile oak (*Q. petraea*) and red oak (*Q. rubra*), are suitable, as well as white oak (*Q. alba*).

129

Oak bark

Preparation

For one hectare use 50 g of oak bark. Decoct for 15 to 20 minutes in 3.5 l of water, and dilute in 35 l before spraying. (Per acre: ¾ oz in 1½ quarts, diluted in 4 gal).

Dandelion tea (*Taraxacum officinalis*)

Dandelion tea reinforces the silica process, improving the quality of plant tissues and their resistance to fungal attacks. This tea is really beneficial to the development of grains, and complements treatments of mixed teas or of copper and sulphur in the spring in vineyard or orchard.

Preparation

For 1 ha (2½ acres) use 10 g (¼ to ½ oz) dried flowers in 3.5 l (1 gal) of water, then dilute with 35 l (10 gal). It is valuable at the beginning of spring, on vines when the leafing stage has begun (4–5 extended leaves).

A dandelion flower at a good stage for harvesting

Camomile tea (*Matricaria recutita*)

Camomile tea (like nettle tea) allows for a reduction in the dose of copper used on vines. Like dandelion tea, it is best used alone or together with small quantities of copper and sulphur.

Preparation

For 1 ha (2½ acres) use 10 g (¼ to ½ oz) in 3.5 l (1 gal) of water, then dilute with 35 l (10 gal).

It has a favourable effect on vines suffering from chronic dryness (Syrah or Shiraz, for example) and those having difficulty getting to maturity. In this case, dilute the base tea in a large volume of water (more than 180 l, 50 gal). One or two passes of this tea at higher concentration (5 to 6 times as much camomile as above) late at night in July or August have also given good results for this particular problem (confirmed by Pierre Guibal, a wine grower in the Herault region of France).

Camomile

Alder buckthorn decoction (*Rhamnus frangula*)

Researchers at Agroscope in Switzerland have found that alder buckthorn decoction can stimulate stilbene production in vines, which is effective against mildew for up to ten days. This effect is only seen when mildew is present.

Preparation

This recipe is by Jean Christophe Pelerin, a biodynamic winemaker in Bugey. For one hectare, used 120 g of alder buckthorn (1¾ oz per acre). Decoct for 30 minutes. Add to horsetail decoction (120 g/ha, 1¾ oz/acre) and meadowsweet tea (250 g/ha, 3½ oz/acre). Copper (250 g/ha, 3½ /acre) or wetable sulphur can also be added if needed.

Spray the whole lot at 120 l/ha (13 gal/acre). An infusion of mountain savory can also be beneficially added at 250 g/ha (3½ oz/acre), or 5% plant extracts such as nettle, comfrey or a mix of the two.

Absinthe

Wormwood decoction (*Artemisia absinthum*)

Wormwood decoction is a repellent against several insect pests, in particular the codling moth, cabbage white, cabbage moth, flea beetles, aphids (on black beans and green beans), and the carrot fly.

Spray two to three times at two-day intervals. It can also be used in vineyards invaded by butterflies or leafhoppers. Spread over the soil for use as a slug repellent. It has a fungicide action on gooseberry rust (as indicated by Eric Petiot).

Preparation

For 1 ha (2½ acres) decoct 100 g (3½ oz) of dried plant or 300 g (10 oz) of fresh plant matter in 3.5 l (1 gal) of water for 5 minutes, then dilute with 35 l (10 gal). This plant can also be used as a fermented extract.

133

Horseradish

Horseradish tea (*Armoracia rusticana*)

Horseradish tea is effective against brown rot in stone fruit trees during the flowering season. It can also be effective against seedling blight (damping off).

Preparation

Make an infusion with 300 g (3½ oz) of roots and leaves for 10 l (2½ gal) of water. It is used undiluted. There is a horseradish extract against common bunt (Tillecur from Dr Schaette GmbH), that is used by pre-soaking seeds.

Tansy

Tansy tea or decoction (Tanacetum vulgare)

Tansy tea or decoction is a good repellent against various animal pests (cabbage white butterflies, cutworms, etc.).

Preparation

Infuse 150 g (5 oz) of dried plants for 3.5 l (1 gal) water. Let it steep for a few minutes, then dilute it in 35 l (10 gal) for 1 ha (2½ acres). You can also use this as as a fermented tea or decoction.

Teas of fruit tree leaves and fruit

Rudolf Steiner indicated that teas from fruit tree leaves and fruit strengthen the horn manure preparation (500) used in orchards or on fruit trees. The tea is mixed in with the horn manure water before stirring for an hour. Harald Kabisch, a pioneer consultant for biodynamics, recommended this tea mixed with the horn manure preparation to help energise old or weak trees. He suggests spraying it abundantly onto soil and foliage.

Rhubarb leaf tea (*Rheum undulatum*)

Rhubarb leaf tea, diluted at 5%, can prevent potato blight. It also has a repelling action against aphids, caterpillars, slugs and leek moths. Rhubarb can be used as a ferment.

Preparation

Put 2 kg (4½ lb) of leaves in 10 l (1½ gal) of hot water and leave for 24 hours. Use undiluted.

Rhubarb

Calendula

Calendula tea (*Calendula officinalis*)

Use fresh, dried or powdered calendula. The parts of the plant used are the leaves and flowers. This controls aphids and brings general health to the plants.

Preparation

For 1 hectare (2½ acres) use 100 g (3½ oz) of dried plant in 35–50 l (9–13 gal) water.

Comfrey tea (*Symphytum officinale*)

Comfrey tea has similar properties to comfrey ferment (see chapter *Ferments*). It is an insecticide and fertiliser and is useful against mildew in viticulture.

Comfrey

Chive tea (*Allium schoenoprasum*)

Chive tea is used against scab on apple trees. Use fresh young chives that have not flowered, add to boiling water and let it infuse for 15 minutes. Dilute this tea 2 or 3 times, and then treat the apple trees. Alternatively plant chives at the base of the trees in the orchard.

Chives

Onion and garlic

Extracts made from onion or garlic are very effective and can be used as preventions against fungal disease and as general insect repellents. They disrupt the behaviour of several pests and insects in the garden. They prevent peach leaf curl. You can also add peelings or scraps of onions and garlic to other plant extracts.

The essential oil of common or wild garlic (*Allium ursinum*) helps to control insects. When faced with insect problems on trees or vines, add several drops to sulphur, clay or other products for spraying. When working with essential oils in liquids, it is advisable to premix it with a little milk.

Note, the aroma from this essential oil is powerful and persistent.

Onion and garlic decoction

This recipe is used in the biodynamic garden of Agrilatina in Italy and has also been successfully used by others. It is effective against aphids and bugs but has no effect on ladybugs (ladybirds).

Add 1.5 kg (3 lb) of onion and 1 kg (2 lb) of garlic to 50 l (13 gal) boiling water. Let it boil for 15 minutes. Spray with a fine mist directly on the crops after it has cooled, without diluting. Repeat this every 2 or 3 days, especially after rain or watering through irrigation.

Simple fungicide preparation

This tea helps prevent and cures Septoria and rust, strawberry rot and several other fungal diseases. On celery, chard and parsley, it has been found (by Rémi Picot, a market gardener in Alsace, France) to consistently protect new leaf shoots.

Preparation

Stir 70 g (2½ oz) of pressed garlic in 10 l (1½ gal) of warm water (20°C). Stir for 20 minutes and let it rest for an hour before applying. Spray this towards the evening, with a fine mist. Repeat on three successive evenings, if needed.

Complex insecticidal preparation

These preparations can be used against may bug or cockchafer, wire worms and cutworms. Soak the exposed parts of the plants (mound or bare roots). They are also good repellents against parasitic insects, in particular cabbage moths and aphids. They can be sprayed twice a month in the greenhouse, on lettuce, cabbage, melons, strawberries, etc. alternating with fish meal spray. These repellents also prevent codling moth in orchards. The first recipe is very good against domestic animal fleas and lice.

First recipe

Mix 100 g (3½ oz) of chopped or crushed garlic with 2 tablespoons of medicinal liquid paraffin. Let it soak for 48 hours. Dissolve 8 g (¼ oz) of soft soap in a half litre (quart) of water. Mix, filter and put in a bottle. When using, dilute 1 part in 99 parts water (increase the concentration if necessary).

Second recipe

For twenty days, ferment 1 kg (2 lb) of garlic, or 100 g (3½ oz) of dried or freeze-dried garlic, with 1 g of soy lecithin in 1 litre (quart) of 70% alcohol (ethanol). Just before using, add 150 g of a hydro-alcoholic propolis solution. Dilute it all in 100 l (26 gal) of water and used at a dose of 35–50 l/ha (4–5 gal/acre).

9. Ferments

A ferment involves soaking a solid in water to soften it or break it up, for longer or shorter periods depending on the required properties.

Nettle ferment

Nettle ferment is a powerful and stimulating fertiliser for all plants. Nettle, like comfrey, stimulates the formation of mycorrhizae and rhizobia which allows better assimilation of mineral elements by plants. If needed the ferment can be applied every two weeks on vegetable gardens or on large crops.

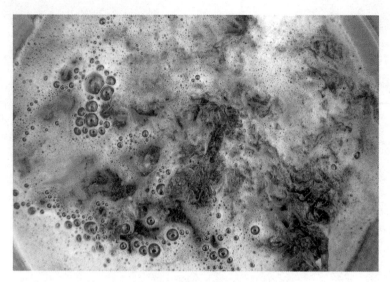

The formation of foam on the surface indicates active fermentation

Preparation

Harvest the leaves and stems of the stinging (or common) nettle plant *(Urtica dioica)* at the beginning of the flowering stage. Adding the roots is a good stimulant for the life of the soil and in particular the formation of mycorrhizae.

Cover 10 kg (22 lb) of fresh nettle or 1.5 kg (3 lb) of dried nettle with 100 l (25 gal) of rain water and let it ferment, stirring occasionally. We should emphasise that the process is fermentation and not putrefaction. While fermenting, the nettles start to float to the surface and may then slowly drift down to the bottom of the vessel. Stir once or twice a day throughout fermentation.

The container is best placed in the shade. Fermentation takes place when a carpet of tiny dense bubbles forms at the surface. If it starts to form only large bubbles when stirring, it is ready for use. This can take from five days to two weeks (rarely longer), depending on season and temperature. Start to use it before the fermentation is complete. If in doubt, it is best to filter too early than too late.

Container with a hinged lid for fermenting

Filter carefully with a stainless steel sieve, cheesecloth or a nylon stocking. After filtering, the ferment can be stored for several months in a barrel, well away from heat sources, light and frost.

The liquid should be used at 5% as a foliar spray and up to 10 to 20% for spraying on the soil.

The smell is powerful, but should never be putrid. Note that it permeates everything that has been in contact with it. So wear gloves! (This recipe is based on my own experience and suggestions by Bernard Bertrand.)

Use

Nettle ferment is effective against chlorosis of the vine. For this, use fermented nettle before the growing season, at a dosage of 20 l/ha (2 gal/acre) to 100 l (25 gal) of liquid, that is, a concentration of 20%. It is also possible to apply it directly

143

Demijohns for storing ferments

to the soil. In case of serious problems, repeat two weeks later (as suggested by Didier Monchovet, a wine grower in Burgundy). Prepared horn manure (500P) can also be used against chlorosis, several times at the same dose or once at a higher dose (200–400 g/ha, 3–6 oz/acre).

Where plants are susceptible to fungal disease, avoid using nettle ferment, or add 3 to 4 litres (1 gallon) of horsetail ferment or decoction.

In viticulture in general, treatment with teas is preferable as they are better against fungal disease.

If you want to use ferments together with copper or sulphur treatments, limit the diluted liquid to 1 to 2% concentration, that is, 2–4 l/ha (1–2 quarts/acre).

For fruit trees and in the vegetable garden, a fresh nettle preparation fermented for 24 to at most 36 hours, is effective against aphids, when applied before the curling of the leaves.

Making nettle ferment

Reinforced nettle ferment

According to Volkmar Lust, this ferment encourages activity and fertility. It stimulates microbial activity in vines and orchards when used on the soil. Foliar sprays are good for grain and for market gardens.

Preparation

Mix 10 kg (22 lb) of fresh nettle or 1.5–2 kg (3–4 lb) of dried nettle, 2 kg (4½ lb) of pigeon or poultry droppings, 1 kg (2 lb) of uncauterised horn shavings, 2 kg (4½ lb) of maerl or dolomite, 2 kg of micronised basalt in 100 l (25 gal) of water, preferably rainwater (but in any case, soft water).

After stirring, hang the six compost preparations from a cross bar (see chapter *Liquid Manure and Animal Bedding*).

Let the ferment sit for a week in the summer, or 10–15 days when cooler. Stir daily to get a good fermentation.

If pressed for time, make a 'prolonged nettle tea'. Simply let the nettles ferment for 24 hours, filter and then introduce the other ingredients, and allow to ferment. The time can be reduced to 4 or 5 days. Dilute to 20% for a solution to spray on the soil, or dilute to 5% for a foliar spray.

Soil spraying

Take 100 l (25 gal) of the base liquid and add 400 l (100 gal) of water to give 500 l (125 gal) of solution. Use 100–200 l of solution per hectare (10–20 gal/acre). Use on the soil as early as March, before budding starts in order to stimulate the microbial activity in vines and orchards where needed.

Foliar sprays

Take 100 l (25 gal) of the base liquid and add water to give 2000 l (500 gal) of solution. Use 200–400 l/ha (20–40 gal/acre). The foliar sprays are good for grain at the end of winter and for market gardens at the beginning of spring. Do not apply too often. Observe the plants carefully to begin to recognise their needs. This will lead to the understanding and experience needed to find the best balance of stimulation, without risk of fungal disease, or poor storage quality.

Comfrey leaf ferment

Both common comfrey (*Symphytum officinale*) and Russian comfrey (*Symphytum x uplandicum*) are used as fertilisers and have medicinal uses. Like nettle, comfrey is very rich in nitrogen and iron, and contains potassium and silica as well as trace minerals like zinc, manganese and boron.

It stimulates the process of potassium in the soil and its use encourages diversity. It can be used if there is a foliar boron deficiency. It can be used to relieve stress after hail or other climatic shock (suggested from analysis by Gérard Augé). Applied just before or around the flowering period of fruit trees, you can also add a few drops of valerian. Applied after the harvest and before the leaves fall, it encourages reconstitution of reserves and enables better bud burst the following spring (according to J.C. Chevalard).

It is made and used like nettle ferment, but is preferably used during the reproductive and fruiting phase of plants. Dilute to between 2 and 20%. Comfrey can also be made as a tea like nettle tea (see chapter *Plant Extracts, Herbal Teas and Decoctions*).

Bracken ferment (*Pteridium aquilinum*)

As an undiluted ferment, common bracken *(Pteridium aquilinum)* can be used against slugs, woolly aphids, leafhoppers and mealy bugs. Diluted 1 to 10 it can be used against simple aphids. It can also be used against wireworms, limiting the damage to potatoes (according to trials done by the Chamber of Agriculture in Gard, France).

The common male fern *(Dryopteris filix-mas)* is effective against woolly aphids, scale insects and slugs.

It is important to harvest the fiddleheads (immature fronds) before they open up their sporangia at the beginning of August. Make it like nettle ferment: 1 kg (2 lb) of fresh leaves or 100 g (3½ oz) of dried leaves in 10 l (2½ gal) of water.

This ferment is used pure or diluted, depending on its use. For tree care, use at least 12 l (3 gal) of fern ferment diluted to make 120 l (30 gal) of liquid to spray; this will cover 1 ha (2½ acres).

This plant is toxic and should be handled carefully. Bracken manure is a fairly powerful insecticide, effective against woolly aphids.

Fern ferment (Dryopteris Filix-mas)

Dryopteris Filix-mas (male fern) is also toxic and was previously used to treat intestinal worms. It has similar properties to bracken and can be used as a ferment. It is effective against woolly aphids, apple aphids, mealy bugs and slugs.

Horsetail ferment (*Equisetum arvense*)

Used as an anti-fungal or in association with other ferments, the horsetail ferment is less effective than the stirred decoction mentioned earlier. It is prepared just like the nettle ferment and used, in general, diluted at 2–20%. It can be reinforced by adding onion peels or fresh chives harvested when flowering.

Alfalfa ferment

Alfalfa ferment is known especially for its richness in boron. It is used on beets diluted at 5% (10 l ferment in 200 l water) at the beginning or during its vegetative phase.

Chive ferment

Chive ferment, sprayed on carrots every two weeks from the beginning of June to the end of July, keeps carrot flies at bay. Take 2 kg (4 lb) of fresh chives and let it ferment in 10 l (2½ gal) of water for ten days, stirring daily.

Tomato leaf ferment

Tomato leaf ferment works against cabbage moth and beetles. It acts as a repellent due to its odour. Repeat this spray every week. This ferment is used also to prevent against fungal diseases of the tomato itself.

Preparation

Crush four good handfuls of fresh tomato leaves (about 1 kg, 2 lb) and infuse in 10 l (2½ gal) of water for several hours to several days. Use undiluted. This ferment does not store well.

Bernd Schimmele fermentation

Bernd Schimmele fermentation, also known as BS Tonne, is a classic biodynamic vegetable extract used to prevent fungal disease and small pests.

Using fresh plants, take 10 g lavender, 5 g melissa or lemon balm, 5 g lovage, 5 g sage, 3 g marjoram, 5 g sorrel and 3 g burnet-saxifrage *(Pimpinella)*; (a small pinch of herbs is about 2 g). Allow it to ferment for ten days in 10 l (2½ gal) of rain water.

Compost tea or ferment

Compost teas are the subject of much research in North and South America, India and Africa. American Elaine Ingham has been pioneering some of this work. They stimulate the microbial activity of the soil and help plants grow harmoniously. They are also an anti-fungal which can be sprayed on seeds and plants (anti-fungal products like Myco-Sin are made by mixing compost extracts with minerals (clay and organic silica) and horsetail).

Preparation and use

Take one part good, young biodynamic compost to 10 to 40 parts rainwater (by volume). Ideally use compost made with cow manure; avoid compost made exclusively from green kitchen and garden waste (a mix of animal and plant compost is best). Let it ferment in a barrel for one to three days depending on temperature. You can add 1–2% molasses, brown sugar, or even ground grain. Other substances can be added like rock powder, wood ashes or various vegetable extracts.

Stir the liquid regularly to aerate it. This aeration by stirring is essential for the quality. Ideally, invest in an aeration pump like those used in fish tanks.

Filter the liquid to remove coarse substances which would block spraying equipment. A mosquito net is about the right fineness since it will still let beneficial microorganisms through. After filtering, the ferment should be diluted to 20% for spraying on plants, or 50% for spraying on soil. This ferment should be used quickly as it does not store well.

In America other substances are sometimes added, such as fish meal, malt barley, whole milk or whey and even yucca plant extract. Some practitioners add bacterial or fungal cultures. Ideally, biodynamic agriculture draws on plant and animal products which are local to the particular farm, as well as local microflora, yeasts, fungi, bacteria and trace elements. Extracts and cultures which originate in different environments should therefore not be used. Existing local microflora and microfauna can be beneficially stimulated by horn manure, compost preprarations, and so on.

Angela Hoffman's compost ferment

Angela Hoffman is responsible for the biodynamic farm at Sekem in Egypt.

Use 50 kg (110 lb) of mature biodynamic compost (2–3 months old in a hot climate) in 2000 l (500 gal) water at 35–37 °C (95–98.6 °F). The compost should be a good mixture of different plant and animals substances; judge the quality by touching and smelling it. Large farms can have their compost analysed to check for harmful bacteria. Finally, add 100 g (3½ oz) baker's yeast and 20 l (quarts) molasses.

10 kg (22 lb) phosphate powder or feldspar powder can be added as needed.

Ferment the whole lot in an oxygenator for 16 to 24 hours and use immediately, or at most 12 hours after the end of fermentation. Only fill the container two-thirds full, since the volume of liquid will increase.

Use in an irrigation system once a week. It can also be sprayed weekly onto foliage or soil, diluted if necessary, at 40–80 l/ha (4–8 gal/acre) for thirsty plants, and every fortnight for crops and pastures.

Weed ferment

Weed ferment, particularly when made from plants that spread through rhizomes and stolons, is good for controlling weeds. Maria Thun advises to stir the ferment on a leaf day, to wait until there is a decomposition of the rhizomes or stolons, and to spray three evenings in a row preferably when the moon is in the constellation of Cancer. (See also the chapter *Pest and Weed Control* for more information on controlling weeds.)

Other ferments

A number of other plants can be used for various problems. Elderberry (*Sambucus nigra*), sage (*Salvia officinalis*), ivy (*Hedera helix*) and greater burdock (*Arctium majus* or *A. lappa*) prevent botrytis or mildew. Some practitioners use boxwood, St John's wort, rockrose, eucalyptus, lavender, citrus tree bark, citrus seeds, aloe vera, bucktooth bark, fleabane, fennel and rhubarb root powder. Several have had surprisingly good and reliable results in bringing general health to plants as well as increasing their understanding of the plant world.

Most of these vegetable extracts can be mixed, though they must be prepared separately as their fermentation and extraction

Elderberry flowers

times are different. A good springtime blend would be nettle, comfrey and horsetail as a spray on the soil. Do not use too much on certain crops (onion, and root vegetables in general) as there is an increased risk of storage disease.

Contemplative observation, and a Goethean science approach to the nature of the plant in its environment, can bring understanding and shed light on some of the secrets of the vegetable kingdom. Books by Wilhelm Pelikan, Frantz Lippert and Markus Sommer on healing and medicinal plants can be profitably consulted.

Treatment in a market garden

10. Products for Stimulating and Regulating Plant Health

The products listed are mineral or plant products conforming to Demeter International standards. All of these substances are listed in the *Codex alimentarius* as permitted products for organic agriculture throughout the world. Some of the substances or commercial products listed are not used or authorised in some countries, so check locally.

Most of the products listed are, in any case, remedies of last resort in cases of diseases or pests (see instead the chapter *Pests and Weed Control*). As we've already seen, prevention by using good agricultural practices is of paramount importance: seeds adapted to the local environment, water management, balancing the humic, calcareous and magnesium characteristics of the soil, using manure and compost from the farm, and other biodynamic methods. Together these are all steps towards balancing the sustainability of the individualised and autonomous whole farm organism in terms of manure and input (feed, seeds, preparations, etc.) and careful use of biodynamic preparations. Working with the rhythms of the day, the (lunar) month and the seasons, and using the preparations, regulatory methods and D8 described earlier, should always be tried before resorting to alternative products.

Minerals

Clay

In the Agriculture Course, Rudolf Steiner emphasises the importance of clay as mediator between silica and lime, allowing cosmic influences to work better. Clay has many uses in agriculture, including as paste or whitewash on trees and vines; added to atomised sulphur to enable a lower dose; blended with herbal teas (in particular nettle) to increase their effect; and in dry or liquid sprays against numerous diseases and pests. It is an essential component of products (like Myco-Sin) that allow a reduction in the use of copper salts in viticulture and arboriculture. (Some attribute the effectiveness of this product to the activity of the aluminium ion, but this is perhaps over-simplistic.)

Some biodynamic practitioners have tried making clay preparations stuffed in cow horns and buried in the ground during summer or winter, or even for the entire year. Use of these clays after stirring has given a variety of results. It is an interesting subject that requires more research.

Most clays are not active pesticides, and they are exempt from regulations governing the use of chemical sprays, etc in agriculture. They act in a purely physical way, creating a protective barrier, and can be used up to harvest time.

The use of clay in animal feed, as well as in liquid manure and animal bedding, opens up interesting perspectives relating to both animal health, and the health of the soil which benefits from improved cow dung.

There are many types of clay. The more common varieties used in agriculture are described below.

Kaolin or white clay is a healing product with excellent covering properties, making it a good, long lasting paste for protecting trees, just as it is. Adding sodium silicate at 1% or 2% enhances its durability. Clays by the name of Surround WP or Argical are calcified kaolin treated to give them greater efficacy. With water added (1000 l to 25–50 kg, 250 gal to 50–100 lb) this type of clay forms a barrier on the leaves and fruit. It can be useful against the codling moth, apple aphids *(Dysaphis plantaginea)* if used as an autumn treatment, olive fruit flies, the pear psylla *(Psylla pyricola),* the vine leafhopper *(Empoasca vitis),* and other garden insects. Treatments are usually frequent, at 7 to 10-day intervals. It can leave a stain on the fruits and vegetables.

Illite is a mica (silicate) clay, rich in potassium with average absorption abilities. Its therapeutic qualities are well known for animals and humans. Its covering and drying characteristics are somewhere between kaolin and bentonite and it can be used in tree pastes.

Bentonite clay, belonging to the montmorillonite-smectite group, has excellent colloidal properties, and absorbs a lot of water, swelling in the process. Variants richer in sodium are more absorbant; other types, richer in calcium, are more commonly used, including in medicine. As it is alkaline based, it should not be mixed with copper products unless the pH of the mixture is less than 7. It is commonly used in powder form in viticulture and arboriculture for its healing properties and for preventing fungal attacks.

Green clay is also rich in montmorillonite and is used therapeutically due to its excellent covering properties. **Lafaure clay** is also in the montmorillonite group. **Velay green clay** is a mix of illite and monmorillonite clay, used in animal feed and against olive fruit flies.

Adding clay – bentonite or montmorillonite – to animal feed at a dose of 50 g (1¾ oz) per livestock unit per day, results in increased health and productivity, as well as increasing the fertility of the manure. This is true for cattle, sheep, pigs and poultry.

Spraying clay towards the end of the autumn on poor, sandy soils (podzols), at a concentrations of 5–10% for spray volumes of 100–200 l (25–50 gal) has given good results for fertility and productivity.

After hail, clay can be sprayed as a complement to valerian. Spray the valerian first, then in the following days spray clay (20–30 kg/ha, 18–27 lb/acre).

Clay is mixed with sulphur and red maerl in the NAB (sulphur, lime and bentonite) preparation (see chapter *Care of Fruit Trees*) and in commercial products to combat powdery mildew and scab. Spraying at 5% (800–1000 l/ha, 85–105 gal/acre) combats peach leaf curl. Spray when the buds just begin to open.

It has also been used against codling moth, psyllid, whitefly and leafhopper.

Depending on purpose, it can be sprayed in a dilution of 1–5%, or as a dry spray at 25–50 kg/ha, 22–44 lb/acre). As a paste it makes a protective tree paint particularly effective against woolly aphid colonies and cankers.

Basalt

Basalt is a very hard mineral which takes years to break down. It doesn't have any active ingredients, but its mere presence in soil can have a regenerating effect. It is rich in silica, iron, calcium and magnesium, and includes many trace elements.

It can bring about spectacular results for vineyards, orchards, and other trees which are struggling. Recommended doses are often too high; in most cases, 200–800 kg/ha (175–700 lbs/acre) is sufficient. It should ideally be used in the autumn, anything from a few handfuls of basalt sand placed into holes made with a crowbar, to the edge of the crown drip, up to the spreading of 1000 kg/ha (900 lb/acre) in difficult circumstances.

Even better is to include basalt in compost; it can be easily added to compost piles (2–5 kg/m^3 (1¼–3 lb/10 cubic feet). For calceous soils during a period of conversion, Volkmar Lust recommends 500–800 kg/ha (440–700 lb/acre) rock powder, preferably very fine (micronised) basalt. He also recommends an annual application for fruit trees and intensive market gardens of 300–400 kg/ha (250–350 lb/acre).

Basalt can also be sprinkled onto animal bedding in barns, 100–300 g/m^2 (3½–10½ oz/10 square foot).

It's effective against pests and disease in orchards and on vegetable crops – use micronised basalt, 30–100 kg/ha (25–80 lb/acre); and against botrytis (grey mould) in strawberry plants – use 1 kg/100 m^2 (2 lb/square foot).

It can also be used in root dip and in pastes for fruit trees and vines.

Maerl

Maerl – a type of red algae coral dredged from sea beds – is a valuable source of calcium, a foliar fertiliser that prevents fungal disease, and a pesticide (against leek worm, potato beetles, snails etc.), complementing horn silica. It can also be added to powdered, sulphur-based treatments. It is different from other calcium mineral deposits in that it is rich in magnesium and trace elements. However, there are environmental considerations when using it, as deposits are dwindling, though the agricultural sector is only responsible for this to a small degree. A combination of dolomite and volcanic rock ash can be used as a replacement.

It is, however, unparalleled in improving the behaviour of plants, and as an insecticide and fungicide. As a dry spray, use 15–50 kg/ha (14–45 lb/acre). It can also be sprayed with water (mix 300–500 g/100 l, 40–65 oz/100 gal). If used as lime avoid annual doses higher than 200–300 kg/ha (180–250 lb/acre).

Diatomaceous earth (kieselguhr or diatomite)

Diatomaceous earth is a sedimentary deposit of diatoms containing silica (90%) and a number of trace elements. It crumbles into a powder that serves as a contact insecticide against white flies, thrips, mites, beetles, bugs, ants, aphids, caterpillars, as well as against slugs and snails. The action is purely mechanical: the powder seals the pores and the insects dehydrate; it is not selective. It therefore does not engender any resistance in the animal.

Diatomaceous earth also has an antifungal property (it is one of the ingredients found in Myco-Sin). It can be used in powder form against weevils in grain stock (use 1–5 kg/t, 2–11 lb/t) and external pests of domestic animals.

For spraying crops, dilute 1–10% in water, or use as a dry spray at 2–10 kg/ha (2–9 lb/acre). Diatomaceous earth is not toxic, and in its amorphous (non-crystalline) form it is not irritating to the lungs (though inhalation of crystalline silica can cause dusty lungs (silicosis), so take appropriate precautions).

The ecological effects of its extraction, however, are considerable. Whenever possible (in particular for weevils) replace it with micronised basalt powder.

Sulphur and sulphur compounds

Sulphur cannot be ignored in viticulture and for growing fruit. It is used against powdery mildew and scab, despite the damage to beneficial insects. Unfortunately, all forms of sulphur on the market are derived from petrochemical products. Biofa sulphur, despite its problematic origin, has some good qualities and is very easy to use. Natural sulphurs (of volcanic or mining origin) are available in some countries.

The powder forms of sulphur are very efficient but are harmful to beneficial insects. The dose can be greatly reduced by mixing with bentonite clay, talc or maerl.

Lime sulphur

Lime sulphur is authorised for use in organic agriculture. It can be used as a fungicide and acaricide (against ticks and mites) during the winter season. It is a very active and is harmful to beneficial insects (both adult and larva) that help maintain the ecological balance.

It is very alkaline (pH over 12), and corrosive. Wear goggles and gloves to avoid contact with skin and particularly eyes. If splashing occurs on the skin or the eyes, promptly rinse with lots of water, possibly adding a couple of drops of vinegar. Avoid using metallic sprayers, especially copper, as they will corrode.

Lime sulphur solution can be used on fruit trees, after the leaves fall, if it has been a bad year for fungal diseases (scab, monilia, powdery mildew, peach leaf curl, etc.). It is a multi-purpose preventative against scab and peach leaf curl when used just before bud burst to just before flowering. It controls some pest populations like codling moth and weevils (as a supplementary treatment). It is effective against the cochineal but only in high doses. It also eliminates moss and lichens.

The solution also acts a stimulant to growth. It is valuable on all sensitive young fruit trees and in the vineyard. Spray apricot, peach, almond and nectarine trees on a sunny day, early in the season (February or March). Lime sulphur solution is a good winter treatment for orchards and vineyards affected by scab and powdery mildew. It is also a good preventative treatment of these same diseases during the growth period.

Making lime sulphur

Lime sulphur is made by reacting calcium hydroxide with sulphur.

Commercial lime sulphur is expensive and not as efficient as when freshly made. Home-made solutions can be used immediately after making, and are more active. You can find instructions on the internet; however often the instructions aren't easy to follow.

Here is an easy recipe for a small quantity. Stir 50 g of flowers of sulphur (sublimed sulphur) in some water to form a loose paste. Do the same with 70 g of lime (quicklime is better but more difficult to work with). Mix the sulphur paste with the lime paste in an old metal pot. Heat and bring to the boil, and immediately remove from the heat. Add 10 l water, mix well, filter with a sieve and use the same day (2 oz sulphur, 3 oz lime, 3 gal water).

The smell is distinctly of rotten eggs and is very corrosive. Wear protective goggles and gloves.

NAB (sulphur, calcified seaweed and bentonite)

NAB (from the German *Netzschwefel, Algomin, Bentonite*) is a mix of equal parts of sulphur, lime and bentonite. It is sold commercially in Germany. It can greatly reduce the sulphur doses needed in orchards and reduce the risk of burns during heatwaves or bright weather. It is an excellent preventative against scab.

It is possible to make this blend yourself. Start by mixing the dry powders, then add a little water, making a paste. Gradually add more water diluting to 1% (that is 100 g of mix to 10 l of water, 4 oz to 3 gal). If an atomiser nozzle is used, reduce the concentration to 0.5%. You will need 800–1000 l/ha (85–105 gal/acre), though some practitioners use far less, only 100–200 l/ha (10–20 gal/acre). If spraying is done during the flowering stage, reduce the concentration to 0.6% for the sprayer, and 0.3% for the atomiser nozzle.

Use regularly, every one to two weeks depending on the pressure and rainfall.

Copper salts

Copper salts, and particularly Bordeaux mixture are especially effective against mildew in vineyards and black rot, acting on contact. They are also useful in combatting late-season powdery mildew, grey mould and botrytis. Copper is an important trace element for plant growth; indeed, vines get through 300–500 g/ha of copper per year (4–7 oz/acre). However, it is also a heavy metal which can accumulate to toxic levels in the soil and is harmful to earthworms. Its worth noting that it appears its toxicity can be limited, for example where soil is rich in humus and full of life, in calceous soil, and where biodynamic methods are used.

Because of its toxicity in soil, the use of copper salts (sulphate, hydroxide, oxychloride and cuprous oxide) is limited by Demeter standards in vineyards and orchards to 15 kg/ha of copper metals over a period of five years, with a maximum dosage of 500 g/ha at one pass (13 lb/acre, max 7 oz/acre). Copper salts are not permitted for use as anti-fungal agents in market gardens or large crops in biodynamic agriculture.

Careful use of biodynamic preparations along with the addition of teas (nettle, osier, horsetail, sage, etc.) to the mixture allows a reduction in the dosage by at least half (which equates to 10 to 20 times less than the manufacturer's recommended doses). Using copper in micro-doses of 50–100 g initially, then

250 g and finally, in case of high infestation pressure, 400 g per ha of copper salt, (¾ – 1½ oz, 3½ oz, 5¾ oz per acre) has been successful in almost any situation on biodynamic farms, provided the treatments were properly followed at intervals of 7 to 10 days (depending on the volume of vegetation, leaf growth and precipitation). Systematic addition of teas is essential at these low doses. Note that copper is phytotoxic (damaging to plant growth) at low temperatures (below 10°C, 50°F).

Bordeaux mixture, invented in the Bordeaux region of France where it is known as *bouillie bordelaise*, is copper sulphate neutralised with lime. Commercial versions are 20% copper. This mixture has a long lasting action (still remaining after 20 mm, ¾ in of rain).

Cuivrol is a Bordeaux mixture approved as a fertiliser with added trace elements, in particular zinc. It also appears to help prevent late-season mildew, and has a gentle stimulating effect on vines.

Copper oxychloride is usually less active than Bordeaux solution, but it is also less phytotoxic. It leaves fewer markings on fruits and vegetables.

Copper hydroxide is less phytotoxic but also less persistent than the Bordeaux mixture. Its advantage is the massive and instantaneous freeing up of cuprous ions that act in an acute crisis. It can also be added to Bordeaux mixture as required.

Cuprous oxide (for instance, Nordox) is persistent and resists washing off.

Copper tallates (made with tallic acid) can be added at a dose of 1 l/ha (13½ fl oz/acre; that is, 50g (1¾ oz) of metallic copper).

Copper octanoate is used successfully in Germany and allows a lower dose of copper to be used.

Copper gluconate (for instance, Labicuper) works on foliage by penetrating the leaf systemically (it is less effective on clusters). In association with other forms of copper, it allows a lower dose of copper to be used. It is most effective in warm climates.

Copper hydrogen carbonate is a hydrate of copper carbonate which has a fairly good efficacy, albeit with high doses of metallic copper. It is a powder which can be dry sprayed. Containing 12.5% copper, it is very soluble and adhesive, penetrating the heart of the foliage or clusters, and is not phytotoxic. It is used in viticulture and in market gardens, in powder form at 20–30 kg/ha (18–27 lb/acre).

Other forms of copper salts are available in powder form (for instance, Algocuivre, Solifeuille), containing smaller amounts of copper. In order to tackle serious infestations, they need to be used in a stronger dose.

Research is being conducted on reducing the doses of copper by adding milled flax seed ferments in wine vinegar with copper sulphate ('EEC copper mixture'). Trials in Brazil are very encouraging, and experience from French vineyards using different root stock and in different climates is also promising. More research is needed.

In an average year, try to keep the dose of copper to no more than 1 kg (2 lb), and in a difficult year, to no more than 3 kg (7 lb). In an acute crisis, try copper hydroxide (see above), or the fire remedy below.

Fire Remedy

Matthias Wolff, a biodynamic winegrowing expert, recommends this method. It can be used in extreme cases of mildew, and also on grey rot and brown rot, and powdery mildew. Apply at 500 l/ha (50 gal/acre).

For one hectare, use 300 g metallic copper (usually copper hydroxide); 3–5 kg (7–11 lb) wettable sulphur (or less if possible, depending on the severity); 1% sodium silicate (but remember that it mustn't exceed 5 l/ha (½ gal/acre), and if the weather is dry, reduce the silicate to avoid leaf burn); and leaf fertiliser such as Bioaminosol at 0.5%, that is, a maximum of 2½ l/ha (35 fl oz/

acre) for full foliage (if foliage is only at the eight-leaf stage, for example, limit it to 1½ l/ha (20 fl oz/acre).

When making the mixture, the components must be added in a particular order:

1. Water
2. Copper
3. Sulphur
4. More water
5. Just before the end, add the sodium silicate, pre-mixed with some water
6. Finally, add the Bioaminosol, also mixed with water. Be careful: the Bioaminosol will form silicon on direct contact with the sodium silicate.

Don't use this mixture more than three or four times per year. Pay extra care when the weather is dry. Don't use silicate during flowering. Silicate is incompatible with acidic substances like Mycosin (because it is alkaline). Wash all equipment well after use and be extra careful with tractor windscreens.

This mixture can be used as an effective preventative in doses of 100 g (3½ oz) of copper and 3 kg (7 lb) of sulphur.

Bioaminosol can also be added to other mixtures to stimulate photosynthesis.

Soft soap

If used at the beginning of an invasion, soft soap (traditionally made from linseed oil) is efficient and inexpensive against aphids, whitefly and cabbage aphids. It works against mealybug infestation as well. In order to penetrate their thin chitinous skin and cause death within hours or days of contact, you need to add methylated spirits (wood alcohol). To treat scale insects, use at least twice, about an hour apart.

Use soft soap on its own at concentrations of 2–3%, or mixed with 2–5% methylated spirits (wood alcohol). Dilute in lukewarm water, adding a small amount of rapeseed or sunflower oil to limit the formation of suds.

At a concentration of 0.2%, soft soap increases adhesion of most products and reinforces their action. Avoid mixing with copper-based treatments.

Talc

Talc is a hydrated magnesium silicate. It is used in viticulture as a drying agent and for healing scarring after hail. Its cleansing properties prevent botrytis and acid rot at the end of the season. It controls wasp and bee attacks on ripe fruit. It is valuable for reducing sulphur doses, for which it is mixed in powder form (for this purpose you could also use micronised bentonite).

Slaked lime (calcium hydroxide)

Slaked lime is good for fruit-tree canker (*Nectria galligena*), as an alternative to copper salts. Use as a paste on the cankered areas, or spray.

It was added to the Demeter standards in 2006, but despite its efficacy and absence of toxicity, it hasn't been authorised in Europe since 2007.

Sodium and potassium bicarbonate

Sodium bicarbonate (baking soda) and potassium bicarbonate are used in some European countries to prevent powdery mildew of grapes, apple scab and other fungal disease (black spot and powdery mildew on roses, tomatoes, squash, cucumbers). On acid soils, baking soda or potassium bicarbonate prevents the acidification linked to sulphur additives.

For vines, 6–8 kg/ha (5–7 lb/acre) of baking soda, possibly adding clay, can replace all sulphur-based treatments. Potassium bicarbonate is more costly but more effective. On Mediterranean vines the results are less good.

In the garden, 50 g of baking soda in 4 l (1½ oz in ½ gal) of water gives appreciable results for many fungal problems. To improve adhesion, add a wetting agent or some clay.

Sodium silicate

This simple industrial product is not toxic to the environment, but is very alkaline and thus caustic. It is a skin irritant and can cause serious damage on contact with eyes. It forms a glass film on drying so it is important to wear safety glasses and protect tractor windows during use. Sodium silicate added to tree pastes used for large wounds and canker treatment ensures better adhesion (and is authorised by the Demeter standards as such)

It has many uses, acting on scab, monilia and powdery mildew. It can be beneficial if used immediately after the leaves fall, and again just before the swelling of buds. Use as a spray at 5%. It can be mixed with a clay-based spray of manure tea and whey at 2%.

For monilia problems during the growing period, use at a dose of 0.5% on apple, pear and plum trees, before, during and after the flowering stage. Treat scab at the moment of the swelling of buds, and at the beginning and the end of flowering. Treatments during the swelling of buds are also effective against powdery mildew.

Treat peach and nectarine leaf curl, and coryneum blight on apricots, with a solution of 0.75% sodium silicate (75 g sodium silicate in 10 l water, 1 oz in 1 gal) at the onset of disease. Use at a dose of 0.5% if the leaves are young. Varieties prone to these diseases can first be sprayed with lime sulphur solution when the buds swell.

Some countries do not authorise sodium silicate for agricultural use.

Salt (sodium chloride)

Sea salt or table salt can be used on powdery mildew. Use as a last resort as salt has a negative effect on soil structure.

Dilute in water to 2% and add wettable sulphur. Good results under difficult conditions have been achieved using 10–12 kg in 600 l of water per hectare (9–11 lb in 65 gal per acre). Do not to go over 2% concentration as this can result in significant foliar burns. Table salt has also been used successfully as a curative against confirmed mildew at a dose of 1.5 kg in 100 l of water, further diluted to 500–1000 l per hectare (1½ lb in 12 gal, further diluted to 50–100 gal per acre).

Condy's crystals (potassium permanganate)

Condy's crystals are used against powdery mildew in viticulture as well as against rust and scab, and for disinfecting greenhouses, though it is not allowed in some countries.

It needs to be very diluted, 100–200 g/100 l (½ –1½ tsp/gal), as it is very phytotoxic (damaging to plant growth) at concentrations greater than 300 g/100 l (2 tsp/gal). Do not mix with sulphur. If used on powdery mildew it must be immediately followed 24 hours later by an application of sulphur. For example, for wettable sulphur at 12–15 kg/ha (11–13 lb/acre), or in powder form at 15 kg/ha (13 lb/acre) mixed with the same amount of clay.

In a clay-based tree paste it is very effective against fruit tree canker. Use at a concentration of 1%. In winter treatments, use at concentrations of 10–20 g/l (1¼ – 2½ oz/gal).

Ferramol (ferric phosphate)

Ferramol, a phosphate of iron, is used on large farms, market gardens and backyard gardens as a slug and snail repellent. It is expensive, but it is the only slug repelling product allowed in biodynamic and organic agriculture. Use at a dose of 50 kg/ha (45 lb/acre).

There are many other preventative measures against slugs which should ideally be tried first, in particular horn silica, fermented teas and ferments and incineration.

Organic and herbal products

Milk and whey

Milk, skimmed milk and whey are good against powdery mildew. They can be incorporated into a tree paste for use in the winter months. Diluted to at least 1 to 10, milk or whey can be sprayed effectively in vineyard and gardens, both as a preventative of and a curate for powdery mildew. Plenty of liquid is needed to ensure a good coating. You can also blend it with a wettable sulphur agent. Before flowering, try alternating the two products.

Research in Brazil and Australia has confirmed its efficacy. Skimmed milk also prevents mildew in warmer climates; whole milk is effective too but can clog sprayers. The tree paste is very good as it causes acidification. Acid whey (obtained while making acid cheeses, like cottage cheese) is best, but sweet whey (from making rennet types of hard cheese) has also shown some good results. For tree pastes, you can also add products containing lacto-ferments like kvass or lacto-fermented juices.

Whey can also be added to Bordeaux mixture to reduce the pH.

Essential oils

Essential oil of garlic and fennel are good regulators of both fungal disease and insects. Sylvester pine, eucalyptus, lavender, lavendin, mint, savoury, sage and citronella essential oils can play a role as insecticides and stimulate resistance.

Citrus essential oils, particularly grapefruit, have a beneficial antifungal effect. They are costly, but very active, and should be used sparingly.

Patrice Lescarret, biodynamic winemaker, notes that, on vines, a few drops of rosemary essential oil, diluted in water with a little milk and sprayed onto large pruning scars, has effective drying properties.

Use a maximum of 10 ml/ha (1 tsp/acre); in some cases a few drops will suffice, especially when stirred into plant tea extracts. They complement teas and ferments. They can be used with common copper solutions and sulphur-based products. In market gardening, orchards or vineyards, they complement treatments supporting a balanced environment. They are photosensitive, use them in the evening.

Essential oils need to be diluted in a fatty substance as they cannot be mixed directly with water. For example, emulsify the oils with milk just before stirring into water-based preparations.

Always use good quality essential oils and avoid those which are unethically produced.

Much research is still needed to ascertain exactly which oils are best used for which purposes.

Cade (juniper) oil

This oil, suggested by Daniel Noël, is especially good for disinfecting large pruning wounds. It has a cauterising effect that prevents the spread of fungal spores. Apply undiluted with a paintbrush.

Neem oil

Neem oil is a widely recommended natural vegetable insecticide for tree caterpillars, beetle larva, crickets, leafhoppers and green leafminers. It is not very effective against adult beetles, aphids and whiteflies, and has no effect on scale insects, lice, firebugs, fruit worms or mites. It has a low toxicity level and is not known to be harmful to butterflies, warm-blooded animals and humans. It is very useful in market gardens and for fruit trees.

Pyrethrum extract

Pyrethrum extract is the first choice in the fight against a number of predator insects, especially since rotenone was banned. It works by ingestion, inhalation and contact. Use just before the harvest; it is sensitive to ultraviolet rays, so it is best to use it in the evenings and never when bees are active.

It is mildly toxic to humans and warm-blooded animals, and is highly toxic on all insects and aquatic life, so use with great care. (Just because a product is 'organic' or 'natural' does not mean it is without environmental side-effects.) In most cases, try neem oil.

It is the only product that can be used in organic agriculture for mandatory treatments of the grapevine leafhopper (*Scaphoideus titanus*) that carries flavescence dorée (grapevine yellows). For this particular problem, use essential oils in addition to specific biodynamic treatments (see chapters on *Pest and Weed Control*, and *Viticulture*).

Demeter standards also allow natural pyrethrum for market gardens and large crops for potato beetle, aphids, carrot fly, etc.

Quassia (bitterwood) extract

Extracts from quassia or bitterwood shavings make an insecticide that has been used for a long time and is authorised for use by Demeter standards as a decoction, though it may not be allowed in some countries. It is useful for fruit trees especially on apple and pear sawfly (hoplocampa), on fruit tree leafrollers and aphids.

Spray as a decoction (2–5 kg/ha, 2–5 lb/acre); it can also be mixed with a horsetail decoction. It works on a broad spectrum of fauna, though less than rotenone and pyrethrum. Quassia is not toxic for humans and warm-blooded animals. Like all insecticides, limit its use. The decoction keeps for a whole season.

Terpenes and pine oil

Terpenes derived from pine and mint are surfactants with a dispersing effect. They mix well with fungicides like copper and sulphur, often working in synergy, resulting in an insect repellant. They are widely used in orchards and vineyards but certain varieties of grape (like Sauvignon, for example) are very sensitive to burns caused by terpenes. Terpenes are added as an adhesive agent to vegetable extracts, and to improve keeping qualities.

Terpene products are easily obtained commercially and they are efficient.

You can make a fresh oil yourself by macerating pine, fir or other resin in alcohol. You can also add the cones, chopped in a grinder. Allow to ferment for at least two weeks, stirring daily. It is used as a concentration of ½–2 in 1000 (by volume).

Soapwort *(Saponaria officinalis)* ferment or tea is also a valuable additive to give better adhesion to plant products. Sodium silicate, soft soap, crushed linseed fermented in vinegar, and clay are also adhesives.

Stifenia

Stifenia is an extract of fenugreek which helps stimulate plants' natural defenses. It's a good pre-flowering treatment for powdery mildew on vines, apparently as effective as traditional treatments. It is especially useful on vulnerable vines. It is a systemic treatment and is effective for up to fourteen days. Mix thoroughly and filter before use. Spray 1½–4½ kg/ha (1½–4½ lb/acre); in the latter case, it also acts as a moderate preventative against mildew. Apply from the 2–3 leaf stage until flowering. Some winegrowers have noted its effectiveness even after flowering.

Vegetable and mineral oils

Mineral oils (white oils), although derived from petrochemical products, are good in viticulture or on fruit trees for winter treatments to control a wide range of insect pests, aphids and fungal issues. They work by causing suffocation. Mineral oils do, however contain problematic benzenes. Rapeseed oil is more natural and has the same effect on pests. For spraying, use 1–3 l/ha of paraffin oils diluted with 500–1000 l water (½–1½ qt/acre diluted in 50–100 gal). A number of commercial products are available: read the instructions carefully as they can be incompatible with other products such as sulphur. In many cases, clay-based tree pastes with horsetail and fresh manure are sufficient.

Anthroposophic/homeopathic products

Myco-Sin and Myco-SinVin

Myco-Sin (and Myco-SinVin for viticulture) is a biodynamic product that supports resistance against mildew. It is used successfully in Germany, Austria and Switzerland. In trials at the Research Institute for Organic Agriculture in Switzerland (FiBL), it showed positive anti-fungal effects on vines over the course of an average year. For northern vineyards with high yields, this does not apply: Myco-Sin is useful only at the beginning of the season. Trials in organic agriculture using

Myco-Sin as a replacement for copper on potatoes produced fair results in years when infection pressure was moderate.

This product is made from several clays, horsetail extract, compost extract and diatomaceous earth (a siliceous sedimentary rock known as *kieselguhr*). Care needs to be taken, particularly in the use of copper salts after treatment: at least 15–20 mm (½–¾ in) of rainfall is needed to avoid burning the leaves. Certain varieties are fairly sensitive (Sylvaner and Chasselas) and can undergo a slight burn (read instructions thoroughly before use).

Tillecur

This product comes from Dr Schaette GmbH and was developed from the work of Hartmut Spiess from the Institute for Biodynamic Research in Darmstadt, Germany. It is a product based on horseradish and is very active against wheat bunt. It also has a slight scarecrow effect (that is, keeping birds away).

Thuja D30

Thuja is a homeopathic remedy for many contemporary illnesses including premature ageing. It is effective when the skin, mucous membranes and nervous system are deficient. It is a diuretic and a detoxifier.

This biodynamic treatment was introduced by François Bouchet. It has been used on particularly toxic soil, and on land that does not react well to biodynamic preparations. Use Thuja D30 (thirtieth decimal dilution in homeopathic remedies) at a dose of 35–50 l/ha (4–5 gal/acre), preferably in February or March. Normally, one application is sufficient, followed by a resumption in the normal schedule of biodynamic preparations.

For one hectare, start with a vial of Thuja D20 (available from Weleda). In a clean, sterilised glass flask mix 5 ml of D20 to 45 ml of spring or pure rainwater. Shake the tincture rhythmically and energetically for three minutes. You now have 50 ml of Thuja D21. Repeat this operation successively up to D27, each time adding 5 ml of the previous dilution to 45 ml of water in a fresh, sterilised flask, and shaking for three minutes.

Then take all 50 ml of D27, mix with 450 g of water in a clean water bottle, and shake for three minutes, to obtain 500 ml of D28. Then mix this with 4.5 l of water in a container such as a carboy. After a further three minutes of rhythmic shaking, you have 5 l of D29. Add this to 45 l of water and stir for 3 min. You now have 50 l of D30 that is ready to be sprayed on the fields.

For very large areas which require a greater volume of liquid, start again from D25 or D26. D25 or D26 made with 45% or 70% alcohol will keep for several years and can be easier to work with to reach D30.

In mixing, always follow the formula: one part D x plus nine parts water, shake for three minutes, producing one $D(x + 1)$.

Propolis tincture

Propolis is a resinous mixture collected by bees from tree buds to seal gaps and cracks in the hive.

Used on its own, it has a intense anti-fungal effect which is both preventative and curative. It can be added to teas and decoctions; add 20–60 ml/ha (2–6 tsp/acre). Added to tree pastes it reinforces the efficacy of garlic preparations.

On apple and pear trees, it acts on the codling moth and scab, and prevents peach leaf curl and powdery mildew. It has a synergising effect when used as a blend with sulphur. It gives good results against grey rot (2 ml/l, 1½ tsp/acre).

On olive trees, it appears to complete the action of sodium silicate and limits damage from the olive fly. Use several times starting in June, at a dose of 800 l/ha (85 gal/acre) at a concentration of 0.5%.

According to a recipe of Helmut Kühnemann (in *Gemüse)*,

propolis tincture is made as follows: stir 100 g of propolis and 1 g of soy lecithin (as an emulsifier), in 1 l of rainwater (13 oz propolis, 1 tsp soy lecithin, in 1 gal) for 2 minutes every morning for six days. Filter the solution and pour into a bottle. Shake the residue in 1 l methylated spirits, adding 1 g soy lecithin (shake in 1 gal wood alcohol, adding 1 tsp soy lecithin) for 2 minutes in the morning for 4 or 5 days. Filter and pour this solution in another bottle. Immediately before use mix the two solutions in equal parts to the quantity needed. For 2–3 ha you need about 100–150 ml in 100 l water (for 5–7 acres about 3–5 fl oz in 25 gal).

It is also a good supplement to tree pastes and in the vineyard. A few drops are sufficient; you should be able to smell it.

Biological controls

Bacillus thuringiensis (Bt) and granulose virus

These preparations are based on a bacterium that secretes toxins attacking the digestive tract of certain caterpillars or larval insects. These products are used against caterpillar infestation, in particular codling moths, eudemis moths and cochylis moths. Their use is improved by the addition of clay (5–15 kg/ha, 4–13 lb/acre) and sugar (1 kg/ha, 1 lb/acre). The water used must have a pH between 6 and 6.5. They are sold under various brand names; Novodor targets the potato beetle.

These products are all photosensitive and should be used in the evening; they are effective for 1 to 2 days. They are easily washed away by rain. Adding a sticker agent of 1 l/ha (16 fl oz/acre) improves their adherence. Remember also that Bt and similar products do not work well at temperatures below 15°C (60°F).

Resistance to Bt is becoming more common (as it also is with plant-based insecticides). This shows that these preparations are not a lasting solution. Regeneration of the soil and encouraging local plant varieties are better long term answers.

Pheromones

Authorised by the Demeter standards and used with success in a number of different regions, the use of pheromones is not without difficulties. See 'cluster worms' in chapter *Viticulture*).

11. Pest and Weed Control

Using certain vegetable extracts (see chapter *Plant Extracts, Herbal Teas and Decoctions*) together with the ashes of animal pests and weeds are specific practices of biodynamics. They were suggested in the Agriculture Course by Rudolf Steiner and have been researched and developed by various biodynamic institutes around the world.

For fungal disease, Steiner indicated several methods, in particular the use of horsetail tea, oak bark preparation, and keeping pastures and damp areas where mushrooms can thrive.

For pests in general, Steiner suggested meditating intensely, visualising the pest problem on the farm between mid-January

Colorado beetles on a potato plant

and mid-February. It is helpful to do a little background research on the farm or garden, as well as into the weed or pest in question: into its biology, origin, life cycle, the conditions in which it appeared on the crops and the ways of fighting it. Try to accurately observe the weed or pest, and perhaps draw it on paper. With weeds, study them in the crops where they proliferate, observe and, if you can, draw their root system. Then contemplate these drawings in a meditative way, and consider the balance of the various factors: health and equilibrium of the farm, vitality, psychology, human, economics, etc. Following this, solutions may become clearer.

Goethe claimed that if we contemplate a phenomenon without prejudice, without any preconceived notion, but with the certainty that there *is* order and sense in the world, we can begin to understand the 'language' of this phenomenon, and grasp something of its true nature. That way, we develop intuition of how to improve the situation and find a better balance. Our active interest in a weak plant or sick animal is already an important step towards its healing.

All this may seem a bit of airy-fairy nonsense (or even magic), but in the everyday practice of farmers and gardeners, these approaches have often produced surprising results in regulating weeds and pests.

It is important to be aware that the processes at work in ashing are of a spiritual nature. This can be better grasped through a deeper understanding of Rudolf Steiner's conception of the world. In the ashing process, the moment of combustion provides an opening to the world of the elemental beings or the 'group soul' of the plants or animals that are proliferating in a damaging way. It is a kind of religious act in the true sense of the word (*religare*, to reconnect).

Here again it is the personal factor, that is, an individual relationship or connection with plants, soil, forces and beings of nature, that is crucial.

Incineration and the spreading of the ashes is a regulatory method: it does not eradicate the pest but diminishes its reproductive abilities and, like a repellent, helps keep the pest at bay where a boundary is made with the ashes. It is a slow method that has to be repeated several years in a row. It respects the environment more than using aggressive and destructive repellants, even if they are approved for organic use. It also allows the opportunity to ponder the question of why the invasion of insects, weeds or animals came about in the first place: what circumstances created favourable conditions for such an imbalance? Before simply destroying the invasive pest, we should explore what we can learn from the situation.

Without sensitivity to these issues, any actions are in vain. Remember also that the basics of biodynamics – rotation of crops, general diversity, the use of mature compost, appropriate work of the soil, and intensive use the preparations – are essential.

A bug box (insect hotel)

Generally, this method of control has been found to be good for various pests. However, in certain cases, there have been insignificant or deceptive results that are inexplicable. This is part of the mystery of the relationship between humans and nature.

Invasive plants

Hartmut Spiess from the Institute for Biodynamic Research in Darmstadt, Germany, conducted research both in open fields and in containers, observing the morphological development of weeds and their germinating qualities.

According to this research, the use of ashes gave significant results when prepared and sprayed around full moon. The results were much more convincing if the farm or garden was treated as a whole farm organism as described in the Agriculture Course.

Maria Thun conducted a whole series of experiments that, for the greater part, require more confirmation. She suggested

Wetlands are important for the regulation of fungi

different positions of the moon or planets in the zodiac for preparing and spraying ashes. We shall look at these in greater detail below. She generally used homeopathic potencies of up to 8 dilutions (D8) because they were more useful over large areas – in full knowledge that direct application is more effective.

Results from using ashes made by burning weeds are not consistent. A lot of research is still needed in this area.

Ashing techniques

For pests, use adult insects or vertebrates from the area where the ashing is to be done. For invasive weeds, first adapt proper agriculture practices (rotating crops, working and balancing the soil, using compost, proper drainage, etc.). Only then try ashing the reproductive organs of the weeds (seeds, rhizomes or even roots).

Moths or insects can be caught with conventional glue traps (coloured paper or pheromone traps). Sucking or shaking plants

Incineration and ashing (from Thun, *Gardening for Life*)

also works. For mammals or birds, use a piece of skin – in particular the skin surrounding the neck, between the scapulas. For snails or slugs, use about a hundred.

It is best to make the ash with a portable burner in the location of the invasion. Make a fire preferably using hardwood (beech, olive, old vines). To avoid seeds or insects scattering on contact with the fire, put them inside an egg carton and burn the whole carton (step 1 in diagram above). After completely burning it, gather only the grey, light ashes (avoid the dark charcoal; however, if used as a dry spray, you can use both the wood and pest ashes). Grind the ashes in a mortar for one hour (step 2).

Use of ashes

Before dry spraying, the ashes can be mixed with basalt, maerl, clay, sand, sulphur-flower, etc. depending on availability and the nature of the soil. Spread the ashes manually or with a dry sprayer.

Alternatively, the ash can be used homeopathically (or rather isopathically), by successive dilutions in water and rhythmic agitations up to D8, though some biodynamic practitioners have experimented with dilutions up to D14 that seem to work with ants.

How to dilute to D8

In order to get 500 l of D8 to allow a spraying over 10 ha (25 acres), you need 50 l of D7, 5 l of D6, 500 ml of D5 and 50 ml of D4. (As the dilutions are all in a ration of 1:10 it is easier to use a metric measure for the smaller quantities.)

Start with 1 g of ashes and add 9 ml (cm³) of good quality rainwater. After rhythmic shaking (each person must find their own rhythm and shaking pattern) in a clean bottle for 3 minutes, the result is D1.

Now add the 10 ml of D1 to 90 ml of water and repeat the shaking as before. You now have 100 ml of D2.

Then take only 10 ml of D2 (discard the rest) and mix with 90 ml of water. After rhythmic shaking, you have 100 ml of D3. Take this and add 900 ml of 70% alcohol (or 40–45% fruit alcohol), shaking again, to give 1 l of D4. Store for future use. The alcohol will stabilise the solution and avoid deterioration.

All these first dilutions can be done in a bottle. For quantities of up to 15 l, hang the container with ropes from a beam or a branch so that it can swing freely and allow vigorous shaking without having to support its weight. For even larger quantities, stir in a barrel with your hand or a stick. Or you can buy a mechanical stirrer.

Spray the D8 three evenings in a row as a fine mist.

Incineration of the reproductive organs (seeds and roots) of dock plants can be done at the full moon

Burning times and ashing

Weeds

Rudolf Steiner didn't recommend a specific time for burning weeds, only commenting that burning is in opposition to lunar forces. Some people choose the day of a full moon, and others a new moon. In our experience it is important that there is a free choice, and it is unnecessary to refer to the planting calendar for burning weeds.

However, Maria Thun gives the following indications for burning and ashing weeds:

- 🌱 Moon in Pisces for vetch *(Vicia cracca)*
- 🌱 Moon in Aries for field or wild mustard *(Sinapis arvensis)*, wild radish *(Raphanus raphanistrum)* and red or purple deadnettle *(Lamium purpureum)*
- 🌱 Moon in Taurus for ground-elder *(Aegopodium podagraria)*, hairy chervil *(Chaerophyllum hirsutum)* and cleavers or stickyweed *(Galium aparine)*

- ⚘ Moon in Gemini for false or wild oat *(Avena fatua)*, grasses in general, chickweed *(Stellaria media)* and silky wind grass *(Apera spica-venti)*
- ⚘ Moon in Cancer for buttercup *(Ranunculus)* and creeping vines
- ⚘ Moon in Leo for sorrels *(Rumex)*
- ⚘ Moon in Virgo for creeping thistle *(Cirsium arvense)*, coltsfoot *(Tussilago farfara)*, horsetail *(Equisetum arvense)* and field bindweed *(Convolvulus arvensis)*
- ⚘ Moon in Libra for the gallant soldier *(Galinsoga parviflora)* and sow thistles or hare lettuce *(Sonchus)*
- ⚘ Moon in Scorpio for black nightshade *(Solanum nigrum)*
- ⚘ Moon in Sagittarius for saltbush or orach *(Atriplex)* and couch or quack grass *(Agropyron repens* or *Elytrigia repens)*
- ⚘ Moon in Aquarius for stinkweed or field pennycress *(Thlaspi arvense)*, shepherd's purse *(Capsella bursa-pastoris)* and knotgrass and other polygonums *(Polygonum)*

It has been noted that after working the soil while the moon is in Capricorn, there is little germination of weeds, while work done with the Moon in Leo wakens dormant seeds in the soil.

Mushrooms and fungi

Making ash from pruned branches contaminated with powdery mildew, or leaves attacked by leaf curl at moon's perigee while the sun is in Aquarius, gives promising results when the ash is dispersed in contaminated plots. A slurry of diseased plants with ash made from the rest of the fermented (and dried) plant is also effective.

For apple scab and monilia, Maria Thun advises making ash from the contaminated fruit (after carefully removing the seeds or pit) when the moon is in Scorpio. Make a D8 solution and spray this three days in a row on the entire tree and its surroundings: soil, trunk and crown. This should be repeated after four weeks.

Insects and animals

For controlling insects and animals, we must be aware of various positions of the moon and planets, depending on the nature of the insect or animal.

In the Agriculture Course, Steiner talks about making ash from birds and mammals (rodents, crow family, etc) when Venus is in the constellation of Scorpio. Enigmatically, Steiner mentions that adult insects could be made into ash more towards the constellation of Aquarius, whereas ash from larvae could be made more towards the constellation of Cancer.

According to Maria Thun, the following times are best to make the ash, dilute to D8 and, if possible, spray the ashes.

- Beetles and other shelled animals: when the moon and sun are in Taurus
- Mealy bugs, slugs and snails: when the moon is Cancer and reinforced by Mars in Cancer
- Flies and mosquitoes as well as moths (pieris, grapevine moth, codling moth), aphids and biting insects (as well as locusts and crickets) when the moon, the sun and, if possible, Venus is in Gemini (note that aphids are specific to a given plant but their predators, like ladybugs, hoverflies, etc. are generalists who go for all species of aphids)
- Moths (grape berry moth is a nocturnal species) and mites: moon and sun in Aries (reinforced by Mercury in Aries)

Incineration of slugs can be done when the moon is in Cancer

Aphids are incinerated when the sun and moon and, if possible Venus, are in Gemini

- Spider mites and red spiders: moon and Venus in Aquarius (first try a nettle tea or spraying horn manure)
- Cutworms, moths and mole crickets: sun in Taurus and moon in Scorpio
- Ants: moon in Leo
- Birds and warm-blooded animals (rodents, crows, etc.): when Venus is in Scorpio and the moon in Taurus.

12. Climate Issues

Frosts and cold snaps

The effects of spring frosts can be reduced by spraying valerian preparation (507) the evening before a frost is forecast. It can also be done first thing in the morning before the sun is up. Use 5 ml/ha diluted in 30–35 l of lukewarm water and stirred for 20 minutes (½ tsp/acre in 6½ –7½ gal).

Adding 5 ml valerian to the first spring horn manure (500) or horn silica (501) preparations can also provide some protection against frost. Add at the very start of dynamisation.

If there are large temperature differences between day and night through the season, spray valerian on its own (ideally

189

in the morning), or mix with horn silica 501 from the start of dynamisation, to reduce the stress to plants.

Droughts and heatwaves

In hot, dry summers, yarrow tea (10 g/ha of dry plant), camomile tea (10–50 g/ha of dry plant), or nettle tea (100 g/ha of dry plant) can be sprayed to help refresh the plants. In extreme conditions, spray horn manure (500 or 500P) in the evening or early nighttime to limit blockages. You can also spray a compost preparation (barrel or birch pit preparation). Spray in a mist on the plots, but also around the plots. If you spray 500, 500P or compost manure on foliage over the summer, it is desirable to spray horn silica (501), if possible, before the harvest, to compensate.

Dampness and lack of light

In the case of a damp summer or autumn, or when the light is weak, spray extra horn silica (501) and horsetail decoction.

Hail

After hail it's important to act quickly, but waterlogged ground often isn't passable in a heavy tractor. Spraying valerian preparation (507) (5 ml/ha in 30–35 l lukewarm water (⅓ tsp/ acre in 8–9 gal) and stirred for 20 minutes) with a backpack sprayer helps reduce the stress on plants caused by the impact of hailstones, and the chilling effect. A few drops of arnica tincture could also be added.

If the plants are also being treated for mildew, add valerian to the copper treatment. Nettle or willow tea can also be used in place of valerian.

Dry spray with clay or talc in the days following the hail. One or two passes with horn silica (501), 8 to 10 days apart, can also be beneficial if growth is well advanced.

If the hail is near the end of the season, copper hasn't been shown to be effective. Use horsetail decoction with added valerian instead. It has been observed that valerian sprayed after the veraison stage improves the development of the grapes, reducing dryness in the finished wine. The wood also seems firmer at pruning time.

∼ PART 3 ∼

Specific Practices

13. Seeds

Seeds and plants adapted to the local environment and to organic or biodynamic practices are of great importance. It was a question about the quality of seeds that led to Rudolf Steiner giving the Agriculture Course in 1924, through which an entirely new form of agriculture came about.

The return of heirloom and local seeds is vital in an era of globalisation and where introduction of genetically modified plants is driven by corporate financial interests.

An onion seed head with pollinating bee

Choosing the right seeds

Ideally in biodynamics, local seeds would be selected for a specific farm that is diversified, self sufficient and itself like a living organism. Selection would focus on food quality: maintaining life forces and structure in the food. There are qualitative tests such as chromatography or sensitive crystallisation techniques that show significant differences in the vitality and quality of produce grown from a variety of seeds. Over the last thirty years, new varieties have been developed that adapt well to biodynamic methods. Results have been excellent with grains, as well as with vegetable crops. Some biodynamic varieties are registered in the European Union. The biodynamic preparations and cosmic rhythms work on seeds in such a way that if the seeds are planted or transplanted in another soil or location, they often adapt well to the new environment.

It is important to remember that propagating seeds is an essential part of both biodynamics, and the production of

Vegetable seed crop

high quality food for humans and animals. Different countries have groups of farms and institutes that aim to exchange or sell seeds: contact your local biodynamic association for details.

Pre-soaking seeds with biodynamic preparations

Over a number of years Martha Künzel and Franz Lippert have experimented with pre-soaking seeds with biodynamic preparations.

Pre-soaking allows for quicker germination. It brings out the vitality of the plants and their ability to fight off disease. Their growth is more vigorous and root systems are more developed. They have better resistance to climatic stress, are healthier and often have an increase in yield. The legume family benefits especially, with better formation of root nodules and greater development of the root system. This leads to an increase of hardiness, health and productivity.

Practical considerations

Treatments are done the day before sowing.

Use a brush or backpack sprayer to sprinkle the seed pile, before turning the pile two or three times with a shovel. For larger quantities, a cement mixer can be used, or a plastic barrel with a tight fitting lid can be rolled on the ground. Leave the seeds to sweat in a bag for two to three hours. Then spread them out and dry them sufficiently to go through the seeders.

For small quantities, seeds can be placed in a cloth bag and soaked for 20 minutes, then strained onto a wooden board or paper towel. Alternatively they can be dried with sifted ashes, basalt or maerl.

Tubers and root vegetables should be soaked for two hours or sprayed. Spray potatoes three times over a week.

In general, seeds should be used within two days of soaking. Carrot seeds should be put in a cloth and wrung dry, and planted immediately.

In market gardens these preparations should be sprayed on the seedbed, or where you have just transplanted

Preparation

For 100 kg (220 lb) of grain seeds, use a pinch (2 g) of compost preparations in 2–3 l (quarts) of warm rain water, and stir energetically for 5 minutes. After stirring wait 12 to 24 hours before using.

Put one teaspoon (5 ml) of valerian in a few litres of lukewarm water and stir for 15 minutes, use right away.

For the horn manure preparation, use 10 g/l (1¼–1½ oz per gal) of warm rain water and stir for one hour. For barrel, birch pit and compost manure with nettle preparations, use 25 g per litre (3½ oz per gal) of warm rain water, and stir for 20 minutes.

Plants stimulated by different preparations

As recommended by Franz Lippert and Martha Künzel:

- ⚘ Yarrow (502) stimulates rye, linseed, ryegrass and all grass seeds.
- ⚘ Camomile (503): legumes, brassicas (radish, turnip, cabbage, canola, and mustard), tulips, potatoes and also flaxseed.
- ⚘ Nettle (504): barley and lettuce.
- ⚘ Oak bark (505): oats, lettuce, potatoes and dahlias.
- ⚘ Dandelion (506): carrots, chicory and endives
- ⚘ Valerian (507): wheat, mange beets, sugar beets, flax, corn, potatoes, legumes, leeks, onions, tomatoes, dwarf beans, peppers, cucumbers, spinach and celery. Valerian is valuable in damp climates and combats the tendency towards mildew. For potatoes, soaking with valerian strengthens resistance to disease, degeneration and mildew.
- ⚘ Birch pit preparation, barrel preparation, and compost manure with nettle: increases the yield of carrots, potatoes and beets, when soaked or sprayed on the seeding beds. For carrots and beets, use 1 part preparation to 4 parts rain water and 5 parts whey (see chapter *Products for Stimulating and Regulating Plant Health*) and stir for 5 minutes, let sit for 24 hours, and stir again for 5 minutes before using.
- ⚘ Horn manure and prepared horn manure (500

and 500P) work very well when used to pre-soak spinach seeds, beets and potatoes, especially in dry climates.

🌱 Horn manure (500), camomile (503) and oak bark (505) generally reinforce the calciferous process, resulting in larger produce.

🌱 Horn silica (501), dandelion (506) and valerian (507) reinforce the silica process, and are related to the cosmos, encouraging the plant's vertical growth.

🌱 Camomile (503) and dandelion (506) generally stimulate root formation.

Wider background

Throughout the process of soaking seeds, the farmer has an opportunity to enter into a close relationship with the life condensed into a seed, and to introduce enlivening forces into the male seed. This connection between a human being and a seed is also essential when storing fertile seeds sustainably, adapting them to new localities, or creating new varieties.

Carrot seed crop

The Hopi Indians used to sing to their corn seeds. Hugo Erbe, a German breeder of biodynamic wheat, used to put his seeds under his pillow so they could enter into close contact with the night world. Even some researchers admit that they talk to their plants.

For further reading related to this, see the second lecture of the Agriculture Course, as well as Rudolf Steiner's lectures *Harmony of the Creative World.*

14. Green Manure

The best way of developing the fertility of soil is through long term crop rotation that includes temporary grass leys for at least two years. Grass fields should be well managed, tightly grazed but without overgrazing. The ideal would be to alternate between mowing and grazing, and then spreading manure and haying after the animals have gone through. The use of biodynamic preparations should be intense: 500P when sowing in the spring, and then after every mowing or period of grazing and in the autumn. Apply 501 when growth is at its peak in the spring and in the autumn before pasturing the animals.

Green manure is valuable for improving fertility quickly. It creates diversity and re-establishes a balance in fields where monoculture dominated or rotations were either too short or too simple. It can also allow a break in the sequence of crops, and improve the health of the soil. It protects the soil from winter erosion by stimulating the mineral content of the soil which encourages the formation of vegetative cover. This prevents the washing away, particularly of nitrates, over winter.

It is also very helpful for the regeneration of the soil between two plantings in orchard and vineyard.

Green manure brings together a wide range of plant species, and works intensively with the 500P and 501 preparations. The extra work is amply rewarded by improved and deeper humus development and soil structure.

Sowing of green manure is best in spring, summer or autumn, as soon as there is a six week break without crops. In winter this time has to be increased to twelve weeks or even eighteen weeks.

A few rules for green manure crops

Diversity

Always mix several species of grasses, legumes, brassicas and other plants. For example, mix six different kinds of grains (rye, oats, barley, spelt, wheat, triticale) with a good variety of legumes, and

Mix of seeds for complex green manure

then mix this with even more species of other plants. The greater the diversity of root systems, botanical families, qualities and characteristics of plants, the better the effect.

The root system

The root system is always at its peak at the beginning of flowering. The most vigorous development of the root system is during the growth of the leaves. As soon as flowering or fruiting begins, root formation slows down, or even reverses; the vegetable matter shows a decline in quality, and is less able to be transformed by the bacteria and fungi in the soil. As soon as root development peaks, at the start of flowering, it is best to mow the field, spray prepared horn manure (500P), and turn over the green manure before leaving it to rest or possibly repeating the process with another sowing. The prepared horn manure (500P) dramatically transforms the organic matter in the soil into humus.

Remarkable root development in green manure

Rhythms for sowing and turning

Maria Thun advised that sowing is best done on a root day, but also works on leaf days. The mowing and turning of the green cover crop should be done during a descending moon, in the evening of a root day followed by a spraying of prepared horn manure (500P) or several passes with barrel preparation.

A practical example

The Agrilatina market garden in Italy work with 15 different crops in the first sowing and five new species in a second sowing that immediately follows the turning over of the first. This technique is used both in the greenhouse and on field vegetables. In three years, the need for compost was reduced from 40 to 10 t/ha in very intensive crops. The soil became brown with an increase of humus and the soil structure has improved considerably.

One application of prepared horn manure (500P) is given on the day of the first sowing, which is in July after harvesting lettuce, melons, kohlrabi, etc. The green manure crop is then irrigated during germination and initial intense growth. As soon as the plant is sufficiently developed to burst into growth, the soil is irrigated to saturation (check with a soil gauge that the soil is damp to a depth of 80 cm, 30 in), then sprayed with horn silica (501). Watering is then paused long enough to allow the development of deep roots. Finding the right rhythm for the watering is very important; the plants have to be thirsty before irrigating again! This prevents them from becoming lazy, and favours rooting.

As soon as the plants are sufficiently developed and flowering has begun, everything is mown with a sickle-bar mower. The plants are cut very low, even taking a bit of earth with the cut, if the soil is not too stony. Prepared horn manure (500P) is then immediately sprayed before lightly tilling with a tine harrow.

Vetch as green manure

Fava beans as green manure

A second sowing is then done, consisting of five species only. It is irrigated and sprayed like the first.

Care must be taken with soil that is not living as it will not be able to assimilate all this plant matter; in this case, most or all of the plant matter can be removed. With perennial crops the plant matter or the broken down green cover crops can create competition or even suffocate the plants.

Here are some examples of mixes:

First sowing (70–100 days)

Species	Weight in kg/ha
Rocket (Arugula)	1.5
Alfalfa	5.0
Canola	5.0
Wormwood	0.2
Mustard	7.5
Camomile	0.2
Green beans	5.0
Dry beans	5.0
Italian ryegrass	5.0
Lupin	5.0
Radish	5.0
Sorghum	5.0
Clover	5.0
Vetch	10.0
Chives	0.2
Wild Fennel	1.0
Total	**65.6kg**

Species	Weight in kg/ha
Arugula	1.5
Alfalfa	5.0
Canola	5.0
Wormwood	0.2
Mustard	7.5
Camomile	0.2
Green Beans	5.0
Dry Beans	5.0
Italian ryegrass	5.0
Lupine	5.0
Field Beans	5.0
Sorghum	5.0
Spanish sainfoin	2.0
Vetch	10
Chives	0.2
Wild Fennel	1.0
Chinese radish	5.0
Crimson clover	1.0
Persian clover	2.0
Berseem Clover	2.0
Lentils	2.0
Valerian (officinalis)	1.0
Total	**76.5 kg**

Second planting (60–90 days)

Species	Percentage	Weight in kg/ha
Vetch	33	17
Lupine	37	20
Mustard	10	5
Italian ryegrass	10	5
Sunflowers	10	5
Total	**100%**	**52 kg**

Other blends relative to planting times

Planting April 15 to mid August (except during heat waves		Planting April 15 to August 31		Planting September 1 to October 1	
32 species	%	19 species	%	14 species	%
Spring fava beans	5	Winter fava beans	10	Winter fava beans	12
Chickpeas	4	Sweet pea	5	Sweet pea	7
Common vetch	5	Common vetch	5	Common vetch	5
Hairy vetch	5	Hairy vetch	7	Hairy vetch	7
Red clover	1	Red clover	4	Winter lupines	10
Crimson clover	2	Crimson clover	4	Winter lentils	10
Persian clover	2	Sainfoin	4	Rye	12
Alexander trefoil	2	Winter lupines	7	Winter oats	12
Sainfoin	2	Winter lentils	6	Wheat	12
Birdsfoot (coltsfoot)	2	Rye	10	Winter canola	1
Field beans	4	Winter oats	10	Winter rapeseed	2
Spring lupine	6	Wheat	10	Chives	1
Spring lentils	5	Perennial ryegrass	3	Arugula	1
Spring rye	10	Winter canola	2	Spinach	2
Spring oats	10	Winter rapeseed	2		

Spring wheat	10	Chives	1		
Forage radish	2	Spinach	2		
Chives	1	Phacelia	4		
Arugula	1				
Basil	1				
Chervil	1				
Dill	1				
Fenugreek	3				
Cumin	1				
Valerian	1				
Calendula	2				
Spurry	1				
Cornflower	1				
Fennel	1				
Spinach	2				
Phacelia	4				
Total	**100%**	**Total**	**100%**	**Total**	**100%**

Complex green manure: phacelia, buckwheat, grains and mustard

Matthias Wolff's green manure mix

Matthias Wolff, a German biodynamic consultant, recommends this mix of around thirty plants, which was especially developed for long-term vineyards, gardens or orchards. It brings nitrogen to the soil and flowering is staggered to encourage insect life. It takes at least a year to implement. Plant between February and April in well-worked soil to prevent existing plants springing up (at least two passes are usually needed). Roll (rather than crush), and mulch to stimulate flowering and preserve moisture. Ideally leave in place for 2–3 years before resowing rows left empty; this is a good mix to use for 1–3 years before replantation of a vineyard or orchard.

Use 40 kg/ha of seeds (35 lb/acre), or half that if only sowing alternate rows. The mix is made up of: Alexander trefoil, crimson clover, phacelia, alfalfa, vetch, sweet clover, sainfoin, mignonette, Chinese radish or yellow mustard, buckwheat; various clovers to make up the majority of the mix; and a smattering of diverse plants to make about 10% of the total: coriander, calendula, cornflower, nigella, mallow, borage, dill, serradella, scarlet pimpernel, cumin, yarrow, wild carrot, fennel, kidney vetch, trefoil, plantain, parsley.

Other variants

There are other more traditional varieties of green manure. Sowing seeds into cereal crops when passing the harrow or disc in spring, or when hoeing, gives abundant growth at the end of the summer (for example, mix several clovers like white dwarf, red, and sweet clover, with several kinds of ryegrass, trefoil, and in calcareous soils, sainfoin and alfalfa). This is only possible where spring rains allow for germination and establishment of plants.

A number of different seeds after the harvest can be used;

try to plant immediately after a pass with the harvester since this often leaves the soil slightly moist which helps germination. This moisture will have disappeared a day after harvesting.

Here are some possible mixes:
- ⚘ Italian ryegrass (*Lolium multiflorum*) mixed with crimson clover and vetch (10 kg, 20 kg and 30 kg/ha; 9, 18 and 27 lb/acre).
- ⚘ a mix of vetch and oats (50 and 60 kg/ha, 45 and 55 lb/acre) or vetch and rye (50 and 70 kg/ha; 45 and 65 lb/acre)
- ⚘ a mix of peas, vetch and fava beans (45, 45 and 90 kg/ha; 40, 40 and 80 lb/acre)
- ⚘ a mix of rye, vetch and rapeseed (60, 40 and 2 kg/ha; 55, 36 and 2 lb/acre)

You can also introduce phacelia, buckwheat, mustard, rapeseed, Alexandria clover and many other plants.

The best blends have at least five grain crops, five legumes and a couple of plants from other families. There are also mixes to improve floral diversity and to encourage beneficial fauna and nutrients for bees (see chapter on *Care of Fruit Trees*).

Summary

Green manure has to be treated like a real crop and intense use of biodynamic preparations is required; irrigation can also give an extra boost. The most important thing with green manure crops is root formation and the nutrients this brings to the soil. Turning under is not always necessary: it can be cut to make forage or mulching.

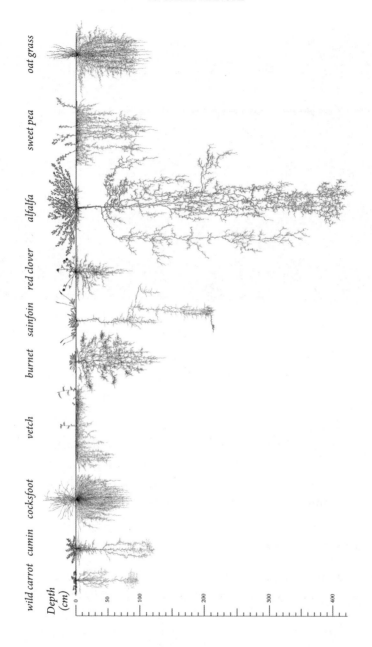

Different root systems, reaching different levels of the soil (from Papapostolu and Thiel, University of Geisemheim)

211

15. Large-scale farming

This chapter discusses the issues specific to large-scale farms which encompass livestock, management of grasslands and large-scale crops. It is intended as a supplement to the general advice elsewhere in this book, not a replacement.

A biodynamic farm

A biodynamic farm should be a diverse, self-sufficient entity. It should have a variety of animals well-adapted to the location (cows, pigs, poultry, sheep, goats, horses, bees, etc) and diverse plant crops (meadows, cereals, vegetables, fruit trees, woods) – not an easy approach in our modern era of specialisation.

However, such an approach can bring superb results, as well as contributing positively to a vibrant rural fabric and harmonious countryside.

We are not nourished by food alone. The air we breathe and our sense perceptions (especially sight and hearing) are essential elements of our health and wellbeing – and this is also true for animals. In fact all living aspects of a farm (including manure, forage, seeds) should be thought of as remedies which can restore the balance of a sick farm – even domestic cats and dogs have their role to play.

Sane agriculture should produce everything that it needs.

Rudolf Steiner, Agriculture Course, second conference

It is necessary to have the number and quality of animals necessary for sufficient manure production. Inversely, we must try to cultivate the plants that the animals instinctively search out to eat.

Rudolf Steiner, Agriculture Course, eighth conference

Livestock

What is it that connects us to our environment, to plants and animals (even domesticated ones)? Aren't animals just, as Backevell described, 'machines used to convert grass and forage into money'? Or as some zoologists describe, 'instruments of which the goal is to produce the most amount of meat, in the shortest time, with a minimum of cost'?

Why, then, do some cultures elevate animals to the level of deities? Many societies have creation myths where animals are honoured and are an important companion – sometimes even a brother – to human beings. And why do we feel such empathy for our domestic pets? Domesticating animals is a natural thing

Gascon pigs

for human beings, literally 'raising them' into ennobled creatures. Cultures which treat animals solely as economic property risk everything. Animals are at the centre of a biodynamic farm because they play an essential role in the fertility of the soil, and the evolution of cultivated plants.

Animal characteristics

Decisions about which animals a farm should have, and how many, should be informed by the characteristics of the animals themselves. Dairy cows, for example, can transform large quantities of worthless forage into protein and fat, as well as providing vital fertilising manure.

Pigs raised outdoors, with a minimum of cereal feed, have the ability to transform waste and the by-products of vegetable, cereal and dairy farming into fat and high-quality meat. Without that transformation, these by-products would contribute to the degradation of the environment (for example, whey in cheese-making).

Flightless poultry are natural consumers of grass and insects, along with fly or mosquito eggs and larvae. It therefore makes sense to rear them to take advantage of these characteristics, and reduces the need for grain feed.

Beware of overproduction, which can often lead to an imbalance in a farm, and can threaten the health and development of the whole. Dairy cows, for example, should only be expected to produce 4–5000 litres (900–1000 gallons) of milk; at this level, they will be content with rough forage, be physiologically healthy, and will produce high-quality milk (recognisable by its qualities when processed and by its impression on the senses).

Respect for physical characteristics

It is essential that cows keep their horns, pig and sheep their tails, and poultry their beaks.

Cows' horns are a key part of the physiology of ruminants, and play an important role in digestion and by extension the quality of their milk and meat. This can be seen in what happens when horns are removed from young cows: the frontal bone becomes porous in compensation, to allow for intense circulation of blood and air in that part of the skull, something normally accomplished by their horns. Without their horns, digestion is affected, their immune system is weakened, the quality of their manure diminishes, their milk loses its structure, and fertility declines.

Cow horns are also essential in biodynamic practice: they are used after the death of the animal in the preparation of horn manure (500) and horn silica (501).

Animal feed

A biodynamic farm should be self-sufficient in animal food (including for pigs and poultry), and should have a strong connection to its physical location.

Leaves, roots and flowering forage are all essential for the health and fertility of a herd. Meadows and rough forage should form the basis of natural food for ruminants. Grain should only be given in very limited quantities, avoiding cereals which could be used for human consumption (bran which is a by-product of flour production could be given to animals, for example). The use of silage should also be limited; in areas where, because of climate issues, the hay harvest is insufficient, haylage is preferred. A direct relationship exists between the quantity of silage eaten, and aggressiveness in a herd; limiting silage intake means that horned livestock can live together safely in an open barn, assuming appropriate living space and feeding troughs.

For dairy cattle, year-round access to dry forage on a self-service basis is important for regulating metabolism.

It's a good idea to permanently (or for a long period of time) allow sections of meadow to grow wild. They will tend towards a natural balance of soil, climate, behaviour and diversity with the livestock. The plants which spring up often have a therapeutic role (and are sometimes called 'nurse plants' (André Ollagnon)).

Frequent use of horn silica (501) will help produce exceptional quality forage, and spraying in spring or autumn also aids animal digestion, which is good for health and production. Autumn spraying is particularly important for obtaining good quality cow dung for use in biodynamic preparations.

Meadows can be mown at flowering stage, or slightly earlier to favour milk production, or later to include ripe grains which are favourable for meat and reproduction. It's important to supplement flowering hay with beets or root vegetables, as well as linseed.

The use of linseed (flax) and roots were recommended by Rudolf Steiner in the Agriculture Course (eighth meeting). Giving soaked linseed to both young and adult animals several weeks before putting out to pasture, and before calving, has an extremely beneficial general effect. The seeds should be blanched to soften and inflate the outer seed coat, to aid digestion. The seed oil is good for hair, horns and especially hooves (few, if any, lesions, resulting in little or no need for hooves to be trimmed).

Roots help to control intestinal parasites and aid strong development of young animals. If root vegetables such as carrots or turnips, or beets, aren't available, compensate by giving bundles of leaves and new shoots in spring which include bark. The cambium and meristem tissues of trees and shrubs achieve the same thing as root vegetables. They can also be dried and preserved for times of difficulty such as illness, growth, lactation and transition periods (return to the barn, going out to pasture). For this reason, a complex environment of hedgerows, groves and woods is beneficial for a farm.

To summarise, a gourmet meal for a cow should include a sprinkling of ground cereal (barley, oats), with pieces of beets or other root vegetables, and a spoonful of blanched linseed. This diet, with flowering hay, will ensure a healthy animal.

Lime salts and minerals

Dr Leo Selinger and Dr Joseph Werr are pioneers of the use of lime salts. The lime salts found in oyster shells, nettles and apatite (phosphate) is good for all animals and especially young animals or pregnant females. For animals being bred for meat, lime salts make a noticeable different to the taste and conservation of the meat.

How to make lime salts

Harvest nettles which are just in flower, dry them and crush into crumbs; keep in a wooden barrel. Crush dry oyster shells to a fine flour in a hammermill. Obtain apatite at concentration D6 (or D3 and dilute yourself to D6 using trituration in lactose powder or bran) or pollen (dried and passed through a coffee grinder). Mix 15–20 litres of dry nettle, 2–3 kg of oyster powder and 100–200 g of pollen or 200 g of D6 apatite (3–4 gallons nettle, 4–6 lb oyster, 3½–7 oz pollen or 7 oz apatite). These rough quantities are then increased by a third to a half by adding wheat bran and mixing well. A three-fingered pinch (about 5 g) of these salts, added to the gourmet meal described above, is enough. This recipe is from André Ollagnon; you may also be able to purchase it pre-made from biodynamic suppliers.

It's good to give this supplement when a farm is transitioning from conventional or organic to biodynamic, at the changing of the seasons (spring and autumn), and during periods of growth, breeding or calving. The three food groups (mineral, apatite; vegetable, nettle; and animal, oyster) together help regulate metabolism, and the supplement is especially good at preventing postparturient hypocalcemia (milk fever) in cattle. It also helps the calf develop a strong skeleton, essential for long life. Dr Leo Selinger recommends that only animals whose mother received lime salts in pregnancy should be retained in a herd.

Mineral supplements can also be made from herbs and medicinal plants, and bundles of spring leaves. This recipe is taken from *The Biodynamic Farm* by Sattler and Wistinghausen (Verlag Ulmer) and has been used, with good effect, at Sattler's Thalhof Farm for a number of years.

A mineral supplement made on the farm supports the animals' metabolism. The dose is 100–150 g/cow/day (3½–5 oz).

- ☙ 200 kg salt
- ☙ 200 kg maerl
- ☙ 100 kg bentonite clay
- ☙ 100 kg wheat bran
- ☙ 100 kg basalt powder
- ☙ 75 kg mixed medicinal plants, comprising 40–50% nettle; 15–25% fennel, tree leaf hay, galega (goat's rue); 35% meadow flowers, dill, camomile, chervil, coriander,

cumin, marjoram, melissa, mint, sage, yarrow flowers, thyme, hyssop; 2% lovage, absinthe, rue

The tree leaf hay is ideally composed of birth, ash, field maple, spruce, hazel, lime and willow. Field maple and lime are good for fertility, while the other species aid vitality and a strong constitution. Harvest the leaf hay before June 24, St John's Day; pick the leaves and tender shoots in the morning. Dry them in the shade or on a ventilated grille, then crush them in a grinder and reduce them to coarse powder in a hammermill. Tree leaf hay can also be added (5%) to cereal feed.

The medicinal plants stimulate vital processes, especially hormonal, due to the vitamins, essential oils and other trace elements they contain. Some medicinal plant mixes are available commercially; check with your local biodynamic association.

Natural medicine

Most biodynamic medicinal and therapeutic research on farms is focused on prevention rather than cure, and is drawn from observing native animals adapted to the land. Non-intensive production and a balanced diet supplemented by biodynamic preparations (especially horn silica 501) contribute to healthy livestock which are resistant to parasites, be it internal (roundworms) or external (warble flies).

If medicine is called for, remedies from aromatherapy, phytotherapy, homeopathy and anthroposophical veterinary medicine are preferred. There are a number of vets who work in these areas.

However, in cases of emergency it's good to know how to administer a few simple remedies drawn from traditional knowledge and based on herbs and plants found in the garden. Particularly useful are cider vinegar, clay, honey, propolis, sodium sulphate, magnesium chloride, cod-liver oil and charcoal.

Amongst vegetables, garlic and mugworth (wormwood) are good against pests, calendula for wounds, and arnica for bumps and bruises.

Plants from the garden or local surroundings can be used to make home-made ointments and ferments with lard, olive oil or beeswax, or teas and tinctures with alcohol or cider vinegar. Teas can be added directly to the forage trough; used regularly, they contribute to animal health, vitality and quality production.

In sandy soils or those lacking in good clay, adding bentonite or montmorillonite clay to animal feed is the best way of rebalancing the whole farm. Keep in mind that not all forms of montmorillonite are suitable (see chapter *Products for Stimulating and Regulating Plant Health*); it's worth doing some trials first; and animals shouldn't drool after absorbing clay. Add it to forage at a dose of 50–100 g (1¾ –3½ oz) per animal per day (suggested by Raimund Remer of Bauk Farm in Germany). This practice should result in good health and productivity, and fertility for the soil which receives the animals' manure. It's a valuable approach for poultry and pigs as well as cows and sheep.

A brief word about salt: some favour rock salt, while others favour sea salt. Animals which originate near the sea benefit from some sea salt (and will also be stimulated by adding herbs to their winter forage).

The health of a herd can be judged by observing its skin and hair, hooves, mucous membranes, horns and dung. Trust the sense impressions you get when first walking into the barn: they're a good indicator of general health. The effectiveness of biodynamic practices in animal husbandry can also be judged by the level of vets fees, the frequency of mastitis and the somatic cell count of the milk, the level of fertility and the average lifespan.

Breeding cattle

Centres of artificial insemination have goals of selection which are geared towards servicing an industry bent on high efficiency and high yields, obtained from a diet of silage and cereals. There is a global trade in semen (for all animals) and a corresponding loss of clear reproductive origin, a reduction in breeding strains and very narrow selection criteria.

In comparison, natural insemination is strongly recommended for biodynamic farms. Biodynamic criteria for selection of breeding animals consider the longevity of the animal, its resistance to disease, its fertility and general health. Animals with strong constitutions, good limbs and good hooves are good candidates, as well as those good with calving and being milked (with an udder with no extra teat). Other essential signs of a good breeding animal are those who favour rough forage, digest well, produce well-formed dung and therefore contribute to the composting process. Good adaptation to the land, level of domesticity, and non-aggressive behaviour are all important factors as well. Selection criteria become even more important when we consider that in biodynamic farming, it

Alpine herd

has been shown that animals can transmit, in a hereditary way, characteristics acquired during their lifetime. This is according to Anet Spengler Neff of FiBL (the Research Institute of Organic Agriculture).

The presence of a bull on a farm, despite the difficulties in handling the animal, seems to respect the natural social relationships within the herd. Natural fertilisation is an important achievement for the female animal after the psychological and physical characteristics of being in heat. The good fertility rate achieved seems to justify this wish to favour natural mounting.

The care of young animals is also very imporant. Farmers should familiarise themselves with the social hierarchies of the herd and develop a strong connection to them – as strong a connection as a calf has with its mother.

Bees

Bees are essential to all agriculture. Their role isn't limited to pollination; they also influence the health of plants and the environment in general. Furthermore the yeasts they leave on flowers aid cow digestion and hence health.

Specific guidelines for biodynamic beekeeping are in the Demeter standards. Bees should be allowed to form beeswax naturally without imposing honeycomb (except as a starter). Natural swarming is encouraged, and honey should be the bees' winter food. Diverse flowering plants should be cultivated, including plants which provide pollen into the autumn and at the end of winter (ivy, willow, dogwood).

Biodynamic preparations and teas should be sprayed regularly around the hive and bottom board. Treat varroa mite with formic acid and essential oils, as well as silica D20 and formica D20. Avoid intervening in the hive when the water element is strong (leaf days, perigee, etc).

The beekeeper should develop a strong internal relationship with the bees and the hive.

Final thoughts

There are many questions these days about the relationship between land areas used to grow food for humans, and land used to grow animal feed. It is inevitable that human beings must learn to consume fewer animal products; however in some areas, especially where crops are vulnerable, animal farming remains indispensable. The future, without doubt, is for farms to be fertile and self-sustaining with as little land devoted to animal production as possible. Many biodynamic farms must continue to work towards this.

We need to explore further local exchanges between farms which, for reasons of soil, sloping terrain, altitude and climate, are oriented towards animal husbandry, and those which focus on plants (vegetables and crops, vineyards and orchards), in order that all can enjoy good levels of fertility, and maintain a responsible carbon footprint.

In the Agriculture Course, Rudolf Steiner said that, in agriculture, we must sometimes loot the earth, but that we must compensate in forces for what we remove in substance.

Grasslands

The secret to keeping meadows and grasslands fertile is to alternate mowing and grazing in a good rhythm.

In many cases, spreading over-intensive manure and mediocre compost results in superficial plant rooting, which will not result in a healthy meadow. If you must bring in compost, apply during the descending August moon, when it will be absorbed well. Another favourable time is after the first cutting – if heat and humidity are good – when it will decompose and assimilate quickly.

Each grazing should be followed by spreading dung, and ideally spraying prepared horn manure (500P), to avoid degradation of the grassland.

Grasslands, like the rest of the farm, should be sprayed with prepared horn manure (500P) in spring, when growth is just starting and the soil is warmed and sufficiently damp. This will stimulate root development and therefore interaction between plant and soil and good grass growth.

When the grass is established, spray horn silica (501) to improve the quality, making hay and pasture more tasty. Ideally, after each grazing or mowing, spraying horn manure (500 or 500P) to provide further growth impulse, then followed again by horn silica (501) once growth is established.

In very dry conditions, spray horn manure (500 or 500P) late in the evening to refresh the aerial parts of the plants.

If a field is harrowed for legume sowing, spraying horn manure 500 or 500P is also extremely beneficial.

We'll finish this short section with a word from André Ollagnon, a farmer in the Rhône:

If you follow all the advice to the letter, the day will quickly become too short. I prioritise by spraying preparations at the equinoxes: prepared horn manure (500P) in the autumn and horn silica (501) in the spring. Then I redouble my efforts on hayfields, cereals, the orchard and the garden, depending on the stage of growth.

Large-scale crop farming

Countryside management

Managing the countryside is a fundamental part of ecology and human health. A healthy balance of forests, meadows, cultivated fields, groves, orchards and gardens has a positive regulating effect on crops and their pests. Wooded areas (including spruce), hedges, flowering borders, diversified fields and natural grasslands are all important. Agroforestry is heading in the right direction.

> We don't live by bread alone. We are also nourished by our sense impressions: images, sounds, smells and ambiance, which we internalise daily, consciously or unconsciously.
> *Thomas van Elsen, FiBL*

It is also recommended to maintain some non-productive areas, not given over to grazing, such as damp fields which mushrooms and fungi can freely grow (which also encourages them to be less aggressive in neighbouring crop fields).

Crop rotation

Long and well-planned rotations are the best way of managing soil fertility, salt levels and pests. Temporary grasslands of two years or more, rich in legumes, are good preparation for a wheat crop. Protein-rich plants such as peas, soya, lentils and beans are the key to good soil fertility. Complex green cover crops are the best guarantee of ecological and economic balance.

Turnips, planted after the harvest, offer a good complement

225

Buckwheat and rye crops in acid soil

to roots for cows and sheep and can be grazed in-situ.

Sattler and Wistinghausen, in their book *Growing Biodynamic Crops*, offer examples of 5-year crop rotations.

Seed varieties

There are several institutes which carry out research into seed varieties and their qualities for biodynamic farming and human consumption. Preserving diversity and understanding local varieties are essential work.

'Improved' seed varieties aren't always better than those that have been adapted to local conditions, especially in connection with reliability and resistance to disease. These 'improved' varieties, along with an excess of soluble fertiliser, often don't develop the necessary symbiotic relationship (mycorrhiza) with fungi, to the extent that they can't even be used as green manure. The quality (strength) of the gluten in these 'improved' varieties is also a contributing factor to the increase in human allergies.

Many farmers are also millers or bakers who are well-placed to evaluate their products and adjust their seed choices as a consequence.

Mixes of seed varieties offer, in general, the best resistance and productivity. Varieties with long stalks can be more resistant to weeds, and produce a good amount of straw for animal bedding. Additionally, such varieties are often rich in silica, giving better general resistance and better taste.

Research is being carried out into cereal varieties which ripen from the bottom up. There is a signficant problem with crops where the grain ripens while the stalk is still green – a false maturity which results in reduced food quality.

In summary, using good local seed varieties is essential and requires expertise. Seeds which are gathered from the farm should be healthy and well sorted.

Manure

As we have seen elsewhere in this book, well-spread manure, along with rich compost, are fundamental to healthy soil, and healthy soil allows plants to draw the nutrition they need. For the most optimal use, observation of the soil and the crops is important. Young (3–4 months) compost can be used on inactive soil; ripe (6–9 months) compost on active soil. Compost is ideally applied to cover crops at the end of the summer, about 15 tonnes/hectare (7 tons/acre). On a meadow which is being turned over for crop planting, 5–10 tonnes/ hectare is beneficial.

Liquid animal manure or reinforced nettle ferment (see chapter *Ferments*) used in spring on new shoots brings vigour to plants and improved yields.

Green cover crops

Green cover crops are useful where crop rotations are insufficient or short, and they bring important organic material which is vital for the structure of the soil. Remember that an important

Grains bordered by trees

part of the work of green cover crops is done below ground; the aerial part of the plant can be grazed, or harvested as an animal feed supplement – it doesn't have to be ploughed under.

On large farms there is a cost to green crops, so they require rigour, careful observation and experience to get the best results. Sowing just after harvest encourages good implantation and gives time for the crop to grow before winter cereals are sown. Don't sow species which can grow rampant, such as vetch or buckwheat. The care taken over ploughing under the cover crop and incorporation is of the utmost importance, and enough time must be allowed between ploughing under and sowing a new crop; use prepared horn manure (500P) after ploughing under to speed up decomposition.

Biodynamic preparations

As is amply documented elsewhere in this book, horn manure (500 or 500P) is sprayed either before or just after sowing, onto damp soil (never onto dust). Don't be tempted to spray at the same time as sowing or working the soil because this doesn't respect the necessary rhythms for the use of the preparations. For autumn crops, spray a second time between the very start of spring and just as the plant starts to bolt.

Horn silica (501) is sprayed during periods of intense growth, and at the start of ripening, if there is time. Depending on the conditions and the crop, horn silica can be used more or less intensively, but should be sprayed at least once.

Horn silica and horn manure have a complementary effect: horn manure can be used to refresh plants suffering from drought (spray onto leaves), while horn silica (which should never be sprayed in dry conditions) is useful when light or heat are lacking.

Example of winter wheat crop

This example is from Benoît Massé, a farmer in Yonne, France.

Horn manure 500 or 500P can be sprayed between rotations, on cover crops or on spontaneous regrowth from September (when the air smells of the undergrowth) or even, always in the evening, during sowing: the same day, the day before, or the day after. If you can't find the time or the conditions aren't favourable, you can do a catch-up when the wheat first appears (if the soil condition is right), or in early spring during weeding or at the moment plants start to bolt, if necessary.

Horn silica (501) can be sprayed in the autumn, at sunrise when the wheat is growing and soil conditions permit, and also if there's a problem with slugs. The ideal time is in spring just after the application of horn manure (500), and then from tillage until bolting. Never spray horn silica (501) in very dry conditions (several days without rain). For more on this, see chapter *Horn Silica (501)*.

Note: the optimal times to spray preparations are often those of peak work on a farm, and perfect spraying conditions (weather, state of the soil, plants, etc.) rarely come along at the ideal time. Spraying requires time and there are often setbacks, breakdowns and unforeseen events. It is therefore advisable to plan for extra manpower at spraying times and have enough spray ready for the surface area.

Working with cosmic rhythms

Rudolf Steiner indicated that cereal crops sown near the winter solstice encouraged a process of regeneration which gives several years of good fertility. Crops sown away from the winter solstice are better nutritionally, however.

If you are able, sowing can be done at favourable times based on the movement of the Moon and planets. For cereals, especially rye, planting before the full Moon can result in better germination and a better yield, and the seeds remain viable for several years afterwards. Before a new Moon, however, the rate of germination and yield are weaker.

For the production of good quality seeds, choosing a good cosmic moment is important. (Other factors are important too, such as rotation – don't sow in depleted soil nor in soil which has just had fresh manure applied. A good soil structure is more important than the level of fertiliser.)

For winter cereal crops, it is often desirable to sow sufficiently early to allow tillage before the start of winter. It's also better to sow lightly to avoid over-rapid growth in the spring when the soil is weak in nitrogen, or the harvested crop will be weak in protein.

Eight to 15 days before sowing, you can use the false seedbed technique with a harrow at a maximum speed of 10 km/hour (6 miles/hour), before repeating just before sowing.

Disease and pests

If slugs are a problem, lightly ploughing the stubble under can help destroy their eggs. Harrowing at night, at speed, on clear soil, will destroy any slugs on the surface. Horn silica (501) can be useful when damage is first noticed, and repeated use of ash or D8 can help in acute cases.

Horsetail decoction at sowing time and around the spring and autumn equinoxes is a good preventative (see chapter *Plant Extracts, Herbal Teas and Decoctions*).

Wheat bunt can cause serious damage. Clover and alfalfa rotation crops can stimulate this disease. A commercial extract of horseradish called Tillecur is quite effective against bunt and also has a slight scarecrow effect (that is, keeping birds away).

A mix of vinegar and Cuivrol – usually used as a leaf fertiliser – at an 18% dose of copper metal also gives good results.

Working the soil

There are many different opinions about how best to work the soil, to what depth, and so on. Much depends on the type of soil and climate, and may also be affected by the cost of energy.

However, acquiring and maintaining good soil structure should be an obsession for any biodynamic farmer. There is an underground architecture which must be respected, and if possible, improved. Smearing reduces the ability of the soil to breathe; only freezing can repair the damage (or alternating humidity and drought, for particular soils).

Never work soil which is too wet and, conversely, avoid working very dry soil since dust can lead to compaction. Effective slow-speed working respects the natural breaklines of the soil, which can improve structure; progressively working with a tine harrow can protect the layers, working to a maximum depth of 5 cm (2 in) in dry conditions.

However, some soils benefit from slightly deeper harrowing (12–15 cm, 5–6 in) with an adapted plough, which also helps to control weeds. Note also that, in very dry conditions in certain contexts, ripping at 20–40 cm (8–16 in) can reduce compaction and aerate the soil deep down.

Philippe Fourmet, a farmer in Meuse, France, recommends the following conditions for producing good quality cereal for breadmaking:

- ⚜ Crop rotation with a suitable preceding crop (grasslands, cultivated grasslands of clover or alfalfa, broad beans).
- ⚜ A mix of modern and heirloom varieties with tall enough stalks to be resistant to weeds.
- ⚜ Sowing a smaller quantity of seeds so they have more light and better nitrogen nutrition.

Summary

For farms which are focused on large-scale crops with few animals, the challenge is how to make the farm self-sufficient. It's important to have a good mix of hedgerows and develop cover crops and green manure, whilst working the soil as superficially as possible and sowing direct. As many different plant species as possible, depending on the soil, climate and aspect, should be introduced into the rotation; ideally legumes, grains and meadow plants should be not less than one third (and ideally a half) of the available land area.

For large farms, applying biodynamic preparations takes time and costs both in material and energy. It is therefore vital that they are applied correctly, for maximum effect on the development of the soil, the quality of the plants, and the quality of food – otherwise, there is a risk of becoming discouraged.

Use of prepared horn manure (500P) can be especially advantageous for large farms – it requires a little more precision than other methods, but is very effective and, even though it requires fewer applications, can give superior results.

Above all, good traditional country knowledge and common sense, and a respect for solid agricultural and organic principles, are the key to good results.

16. Vegetable Farming

The demand for fresh, Demeter-certified biodynamic-quality vegetables is important, be it for local networks (sold in farm shops, markets, etc.) or for national large-scale retail and cooperatives. Growing vegetables in open fields as part of a larger farm is one possible option for diversification.

However, increasing numbers of young, trainee organic and biodynamic farmers are being attracted to more specialised forms of vegetable farming, in cold frames or in polytunnels. It's possible to start a local agricultural business with a small area of land and little capital investment, assuming suitable soil can be found – that is, deep and fertile enough and not requiring too much watering.

Vegetable farming is a career – you can't play at being a vegetable farmer. Professional training is essential to ensure the venture is well set-up, and difficulties and costly mistakes avoided.

Successful farming, of course, is based on general good agricultural practice and paying close attention to crops; good biodynamic techniques will then help to improve the success and quality of the produce.

Planning and preparation of the soil

Vegetable crops, like all crops, need good planning and preparation. Preparation work done in the winter season is the basis of success for a whole crop; for example, the stress of sowing vegetables can be partly relieved by having carefully and visually planned exactly where the seeds will go, keeping a decent gap between crops of the same type.

Similarly, fertilisation must be thought through and planned in advance to get the right balance of fine and rough compost, younger compost, and liquid manure. It's also good to plant vegetables with similar watering needs close to each other.

Rotations between the seven vegetable plant families should be kept as long as possible: four to five years is optimal before returning to the same crop. This strategy can be further improved by using more or less complex green cover crops. Pasquale Falzarano, of biodynamic vegetable farm Agrilatina in Italy, recommends that a succession of two complex green cover crops over a period of three to four months will regenerate and heal the soil as effectively as a longer-lasting rotation. If enough land is available, up to half could be at rest at any one time, either as green manure or meadow. These choices should be made in the calm of the winter season.

Monthly and weekly planning through the seasons and different weather conditions are informed by using a lunar planting calendar which, if used flexibly, can be a useful aid to decision making.

Seeds and transplanting

The choice of biodynamic (or at least organic) seeds is important. Avoid hybrids, even if they are promised to increase standardisation in the final produce, because it will increase dependence on multinational seed companies.

Use young, hardy plants for transplanting. Plants which are transplanted outdoors directly from being under glass can experience shock which affects their future growth and is often a cause of crop failure.

Don't rush: crops can fail if sowing is done too early. Wait for better soil and weather conditions.

Manure and compost

Composted manure which has received the six biodynamic preparations gives better results than commercial organic fertiliser. This is amply demonstrated by the yield, the taste and the conservation of the crop, along with less grass invasion and soil which, after a few years, is easier to work.

In vegetable farming, the decision to switch from commercial organic fertiliser to producing one's own compost is a significant one. It takes a lot of time and energy to source good quality local manure, as well as acquiring the necessary equipment and establishing a good location for the compost pile.

Non-composted organic fertiliser used for sowing or transplanting can cause problems with pests and weeds, but if poorly decomposed manure or lifeless fertilisers cannot be avoided altogether, the effects can be partially offset by also applying prepared horn manure (500P) or barrel compost.

Animal manure (preferably from cows or sheep), even

Spraying horn silica (501) on carrots

in small quantities, makes a big difference to the yield and quality of a crop. Animal products such as droppings, fish meal, powdered feathers, powdered horns, etc can be added to vegetable compost. Compost or composted manure should be used at between 10 and 40 tonnes/hectare (4 to 16 tons/acre) depending on the number and type of vegetable crops. In soil which has at least 3.5% active organic material, this manure will ensure correct nutrition for crops.

It is essential that well-composted manure is used, with no remaining trace of its original material. Composting can take three to twelve months, and all manure should receive the six biodynamic preparations. Air-fermented compost brings beneficial micro-organisms (bacteria, actinomycetes, etc.) which can reduce diseases such as seedling blight (damping off), pythium root rot, rhizoctonia, downy mildew (bremia lactucae), take-all disease, etc. Compost gains qualities favourable to the soil as it matures; ideally, it should mature for at least three to six months. Compost should be incorporated into the soil soon after spreading (not with a plough).

Young compost is useful for leaf vegetables and greedy vegetables such as gourds, tomatoes, etc. Riper compost is good for plants susceptible to fungal diseases.

Compost can be profitably used before sowing green cover crops, or when turning them under (if the soil requires loosening, do this first, when dry). If needed, basalt, maerl and magnesian rocks can be applied in the autumn.

In the growing season, when needed, nettle or comfrey tea or ferment are valuable. Compost teas can also be used regularly both as a fertiliser and to protect plants; use them in the evening just before the full Moon or when the Moon and Saturn are in opposition.

Biodynamic preparations

In intensive vegetable farming with three or four crops per year, the preparations must be applied more frequently: as many as four or five sprayings of horn manure (500 or 500P), and two (spring and autumn) at the very least.

Horn silica (501) is essential for vegetable farming: plants will have a better structure and be more resistant to fungal disease, with improved taste and conservation. Spraying can start on open-air crops in spring, as soon as the plants are sufficiently developed; for maximum effect, spray at the start of growth after the 4-leaf stage, avoiding very young or weak plants. Three applications of horn silica, on the same crop, are often beneficial. The Demeter standards state that all crops must receive at least one application during growth, which is sometimes difficult for crops with short cycles such as radish, cress and cut lettuce.

For crops grown under cover, spraying of horn silica (501) can start earlier because of the relative lack of light and the heightened humidity. If you see well-irrigated and well-fertilised plants losing a little firmness, that's the time for horn silica. At the end of the season, from August 15, re-spray horn silica on vegetables going into storage; for root vegetables, spray in the afternoon 2–4 weeks before harvest.

This spraying of horn silica must be guided by closely observing the plants and the weather. Ensure that plants have enough water, since horn silica can cause a light increase in transpiration.

At the successful Sekem biodynamic farm in Egypt, where the climate is very hot, all irrigated crops receive at least three applications of horn silica.

For more information on horn silica, see chapters *Horn Manure (500 and 500P)* and *Horn Silica (501)*.

Green cover crops

Sowing of green cover crops between two crops or between seasons is very beneficial, leading to a possible reduction by half of the amount of compost needed – this is especially relevant to vegetable farmers, who often have to buy nearly all their manure. Green cover crops with good rotation also helps to control weeds.

The main times to sow green cover crops are in spring and at the start of autumn – but sometimes in summer as well (see chapter *Green Manure*). Winter cereal crops work well but legumes (clover, vetch, winter field beans) are best. For summer or autumn cover crops, buckwheat and phacelia (especially for cereals) develop quickly. If a lot of crucifers are already grown, especially in acidic soil, avoid seed mixes which are based on mustard, rapeseed or rape. The multi-species mixes used by Italian farmers give excellent results.

Green cover crops can be successfully sown in early autumn when the Moon and Saturn are in opposition.

Superficial turning under of green manure should be done just before flowering, and prepared horn manure (500P) should be applied immediately afterwards. The soil should remain in clumps to assist positive aerobic activity.

Excessive ploughing of green or non-decomposed material can be toxic for soil which is not living, especially if horn manure (500 or 500P) or barrel compost is not used. Using a plough or disk harrow to turn under can seem quick and practical but can bring a multitude of problems such as proliferation of click beetles and cutworms, and weeds. It's better to collect the debris and compost it with straw and organic animal products.

The remains of green beans, for example, after the harvest can be broken up in place and treated like green manure. The plot can then be reused almost immediately for transplanted lettuce.

It should usually be three to five weeks after turning under green manure before a new crop is planted; and the delay should be longer for direct sowing than for transplanting.

Fungal diseases and pests

Preventative measures are more important than cures. Biodynamic vegetable farmers shouldn't be trying to combat this disease or that pest; rather, they should be undertaking measures to stimulate the natural immunity and health of their plants – such as ensuring fertile, well-structured soil and frequent use of biodynamic preparations.

Products containing copper are banned in biodynamic vegetable farming, so the use of horsetail decoction, horn silica (501), valerian and herbal teas is essential for vulnerable crops such as potatoes, tomatoes, gourds, etc. Repeatedly spray horsetail decoction onto the soil in autumn and in spring, and even during the crop season, for the best results.

Many vegetable farmers use a mix of herbal teas which, sprayed regularly, are a good preventative (see chapter *Plant Extracts, Herbal Teas and Decoctions* for proven recipes of nettle-horsetail, nettle-yarrow, preparations based on garlic, etc.). Alternating week by week can also be effective.

Regular superficial harrowing can help counteract carrot fly.

Flower border on a vegetable farm

Frequent light watering is effective against leaf beetles and red spiders.

Crops grown under glass and in the winter benefit from a fine layer of silicon sand sprinkled onto the soil, which reflects sunlight and improves plant health. Horn silica (501) can help plants have a more upright bearing, so fewer leaves are in contact with the soil and are better exposed to the light.

Aeration is very important for crops grown under glass, especially if watering is done by spraying: it's better to aerate too much than too little.

Choosing appropriate varieties, from good quality nurseries, will help stave off disease. As a summary, here are the preventative measures recommended by Roger Raffin:

- Thoroughly clear all crop residues.
- Use long rotations (at least 3 and preferably 5 years) and plant diverse green cover crops in between.
- Consider clearing the soil of pests through summer solarisation.

- ψ Water in the morning and aerate carefully.
- ψ Limit the density of plants.
- ψ Ensure good drainage.
- ψ Avoid an excess of nitrogen, including non-decomposed organic matter.
- ψ Prepare the soil thoroughly.

Pests can also be effectively controlled by maintaining varied areas of wild and flowering plants around the crops, providing a home for helpful fauna. Even in a small market garden, a damp area or a pond allows mushrooms to grow in a controlled area without invading the neighbouring crops.

Hedges and windbreaks are good protection again strong winds and help regulate heat and humidity. The biological diversity of hedgerows can also help regulate pests.

Working with cosmic rhythms

Successful vegetable farming is dependent on many different aspects, so don't try to follow the indications in lunar calendars too strictly, especially if they clash with a particular agricultural activity. Gradually increase the use of the calendar, for example, starting with avoiding negative days, especially lunar and planetary nodes. In general, the few days before the full Moon and perigee are good for seed vitality.

Some plants react quite clearly to the sidereal rhythm of the Moon (root days, leaf days, flower days and fruit days). For example, red radishes sown on root days (and secondarily on leaf days) do best. Kohlrabi, spinach and lettuce sown or tended on leaf days can show improvement in consistency of produce. Transplanting is best done during a descending Moon (referred to as a 'transplanting time' in lunar calendars) – especially bare-root transplanting.

Where possible, leaf vegetables and fruits are better harvested in the morning, and root vegetables better in the afternoon. For root vegetables going into winter storage, avoid harvesting on days connected with water (leaf days, full Moon, perigee); descending Moon times are preferable.

Vegetables such as carrots, turnips, radishes and lettuces which are being sown under glass between October and January should be sown during a waning Moon, which helps limit their vitality and stops the seedlings getting too leggy (thanks to Jean Michel Potiron, a farmer near Nantes, France for this suggestion).

To work correctly with cosmic rhythms, it's important for farmers to develop their own experience and sensitivity, and to do simple trials based on the calendar indications to find out what works best.

Working the soil

To work the soil well, it is essential to understand one's own soils, and to have appropriate equipment. Even then, finding the right tool, in the right place, at the right time, isn't always easy.

A rotary tiller may seem like an attractive option but it often leaves a smooth compacted layer behind – so it is better to use a cultivator with teeth (tines). Soil is best worked late in the season, before winter, for spring planting. Be careful of mechanised equipment; and also of working soil which is too damp or too dry; or of harrowing at too fast a speed.

Each time the soil is worked, cosmic influences are at work on the soil at that moment, so it's important to avoid unfavourable times, especially nodes.

Here are Roger Raffin's top recommendations for working the soil successfully:

🌱 Don't dig or harrow deeper than 25 cm (10 in).
🌱 Break up any compacted layers of soil.

Worked soil of covered crops

- ⚘ Avoid rotavators or disc harrows.
- ⚘ Choose tools with teeth (tines).
- ⚘ Choose machines with large wheels to spread the pressure.
- ⚘ Limit the speed of harrowing to protect the soil.
- ⚘ Avoid very fine surface soil; don't churn it up.
- ⚘ Always try to achieve a good, loose soil structure.

One interesting development is the use of permanent crop beds, with permanent wheel tracks either side of the bed (a technique developed by Wenz and Mussler in Germany). This helps prevent compaction of the soil in the beds themselves, and favours the use of tine harrows.

244

Weeds

Hoeing is a good way to control weeds. It can be done with a harrow, and lightly working the soil this way improves plant immunity, provides nitrogen and significantly reduces water evaporation from capillary channels. Received wisdom says that hoeing is worth two sprayings. There are hoes adapted for use with tractors in large fields as well as manual hoes which give good service to small-scale vegetable gardeners.

Stefan Funke, a German biodynamicist, has developed a mulching technique with fresh grass and forage (vetch). The use of vegetable mulch under glass, or on legumes in open fields, combined with biodynamic preparations, gives exceptional results for the growth and health of plants. Weed control and water usage are also improved.

Gas flame (thermal) weed control can be used with carrots, onions, etc. It is a localised technique which requires delicate and precise work, and should be reserved for species which can cope with sudden heat, mainly alliums: leeks, garlic, onions, etc.

Covered crops without artifical mulching

Amongst the many tools which can be used for weed control, the tine harrow with small diameter teeth is useful but must be used prudently. Intervention during the growing season is reserved for hardy species such as potatoes, leeks, cabbages, etc. The tine harrow can also be useful for destroying the weeds in a stale seed bed.

A stale seed bed is an ancient, effective and cheap technique for controlling weeds which involves preparing a seed bed in advance and giving weeds a chance to germinate before destroying them before the crop is sown. There must be enough time available between crops, and it requires precision of depth of working: when destroying the weeds, care must be taken not to disturb seeds further down.

Prepare the seed bed and irrigate if there's no rain. In the spring, if the soil is cold, cover with an agricultural fleece (P17, for example). When the weeds have sprouted, destroy them with a tine harrow. Where possible prepare the seed bed when the Moon is in Leo, to better awaken the weeds; harrow when the Moon is in Capricorn, to suppress weed re-growth.

Many vegetable farmers, even on biodynamic farms, control weeds by using protective covers, fleece or floating row covers which are good against beetles, flies, thrips, moths, etc. or by frequent hoeing. Mulch covers which are biodegradable are more expensive but are closer to the natural thing; floating row covers are very durable and last several years – they can be used in certain circumstances, if absolutely necessary.

Solarisation – mulching the soil then covering, to trap solar energy which heats and kills pests – also requires synthetic material. As well as controlling pests, it can also reduce fungal pathogens in the soil. It's not recommended to solarise every year – instead, alternate with a green cover crop.

Summary

Alain Regnault, a vegetable farmer in the Auvergne, France, writes:

> For me, practicing biodynamics isn't just strictly applying good quality biodynamic preparations to the letter. It's also thinking about crop rotations, which are difficult in vegetable farming, along with compost, green cover crops, caring for the environment, hedges, flowers and meadows. Finally, it's important to continually observe the crops.

Michael Leclaire sums up:

> Regularly reading Rudolf Steiner's Agriculture Course is an invaluable help in increasing one's biodynamic understanding and practice.

17. Care of Fruit Trees

This chapter is primarily for those managing fruit trees as part of a larger farm, rather than professional or amateur orchard specialists.

Advice on planting

Choosing the site

Shallow soil is not suitable for planting fruit trees. In this case, create terraces and bring in good soil so that the roots can have a sufficient space to explore.

The ground must not be too dry, unless irrigation is possible, but soil that is too humid is not good either; in this case, drainage is essential. Areas where groundwater comes close to the surface during winter are unhealthy. Choosing soil that is both deep enough and has good drainage is vital. Choose a sunny spot with a good orientation, and plan sufficient space between trees so that light and air can circulate freely. Trees need enough space to develop freely to type, without which they will be susceptible to disease.

Good humus content and porosity in the soil are essential; biodynamic practices can help to improve them further.

Selecting different varieties

There are a great number of species – just have a look at commercial tree catalogues. There are traditional varieties that deserve to be preserved, but many modern varieties are tasty and are hardy.

Here are a few basic rules:

- ⚜ Observe what is being grown in your region. Ask about tree quality and defects. Meet with neighbouring fruit tree owners, organic producers at local markets, local amateur associations and possibly nurseries. Catalogues of organic and biodynamic tree nurseries are useful references, though a dialogue with the growers is invaluable.
- ⚜ Research the type of tree and locally adapted varieties that suit the terroir, which need little or no treatment against diseases like scab, codling moth, etc.
- ⚜ Plant what tastes good, but take care as some of the tastiest varieties are also the most sensitive, and many cannot be cultivated from hardy rootstock such as M106 nor are suitable as large free-form trees.
- ⚜ Choose varieties to provide a staggered harvest, so as to

have fruit for the whole season. Note that some varieties alternate, that is, only produce fruit every other year.

꙼ Plant a carefully-chosen variety of species in the same orchard to ensure biodiversity and a more stable ecosystem. This system was suggested by Volkmar Lust and confirmed by the Research Institute for Organic Agriculture in Switzerland (FiBL).

Planting hedges and environmental measures

Planting diverse hedges, and creating flower borders as a refuge for birds and insects, help develop a healthy orchard. Roosts for birds of prey and nesting boxes for smaller birds also help.

Bees are an integral part of any agricultural or orchard project. You can buy ready-made blends of flowers and bee forage. (For instance, between mid April and mid May, plant 25 kg bee fallow and 1 kg field flowers per ha (22 + 1 lb/acre); or seven species totalling 25 kg/ha (22 lb/acre) – sheep fescue 30%, red fescue 30%, birdsfoot trefoil 10%, sweet clover 10%, red clover 10%,

An orchard is an integral part of a biodynamic farm

Grassed orchard with roost for bird of prey

white dwarf clover 5%, and phacelia 5%; or the following seven species: calendula, cosmos, cornflower, blue flax, pasque flower, cheiranthus, allionia, nigella.

Poultry can help regulate orchards by scratching and pecking while feeding, virtually eliminate a good number of pests. Other animals, circumstance allowing, help bring about a well balanced and healthy whole-farm organism.

Form and rootstock

The simplest form for amateur growers is the most natural symmetrical form, with a central trunk, and minimal pruning.

For amateurs and as part of a wider farm environment: deep and strong roots guarantee good health and resistance to pests, including field mice attacking the roots themselves. Weak and dwarfed rootstock predisposes the tree to disease. These rootstock lead to earlier fruit setting, but the lifespan of the trees is shorter. Such trees are also more sensitive to climatic stress (drought, heat wave or sudden temperature changes). These

Only professionals should attempt to work with weak rootstock

trees are often used for economic reasons but they require more intensive care and irrigation.

Preparing the location

Thorough preparation of the soil at the point of transplanting is the key to a successful fruit tree orchard. The soil must be well structured and be porous. Good drainage of the soil is vital.

Before planting as an orchard, the land should be pasture with a complex flora, regularly grazed or mowed, and be regularly treated with biodynamic preparations. The area should have grain or green cover crops for at least one year before planting, to ensure a good development of organic matter and limit field mice. A succession of diverse green cover crops, well mowed and partially turned under with a tine tool, regular intensive use of horn manure and horn silica (as indicated in the chapter *Green Manure*) are also good preconditions to a successful orchard.

Avoid turning over the soil too deeply with the plough, thus burying good live soil. Instead, use a tool with tines. Subsoiling can be required when the soil is sufficiently dry to favour breaking up and avoid smearing. However, it is the root development of the green cover crops that ultimately ensures the enlivening of the soil to a depth.

Soil analysis can be useful if deficiencies have affected preceding crops. Knowledge of calcium/magnesium levels and of the pH is indispensable to correct the situation. In case of a calcium deficiency, use maerl in moderate doses. In case of a calcium/magnesium deficiency, try dolomite, whereas if the soil is calcareous and deficient in magnesium, kieserite is preferable.

Use compost or green manure in the autumn (10–40 t/ha, 4–16 t/acre). Spread superficially, or preferably spread it on the previous crop before turning it in.

Grass in orchards

Grass in an orchard is indispensable for soil health and provides material for mulching under trees. Mulching helps maintain an intense life in the soil. Don't let the grass get too old; regenerate it by scarifying from time to time. In orchards where there are field mice, avoid covering during the winter and take special measures such as trapping and dry spraying.

A variety of flowers with different blooming times provide nectar and pollen for beneficial insects that help control pests and diseases. Choose a wide variety: tall and short grasses, legumes, medicinal and aromatic plants.

Species to plant
- Grasses: meadow grass, fescue, perennial ryegrass, orchard grass in small quantities.
- Legumes: clovers, alfalfas, sainfoin, birdsfoot trefoil, trefoil, white, sweet clover.

A grassed orchard

🌾 Aromatic plants and flowers blooming at different times: wild chervil, caraway, parsnips, daisies, cornflower, camomile, yarrow, salsify, marigolds, cornflowers, tansy, St John's wort, fennel, wild carrot, chicory, cosmos, etc.

To maintain these flowering borders, avoid using shredders; use a cutter bar or a scythe to minimise the impact on insects and bees. For the same reason, mow in the early morning or late evening. Take special care of the soil in the springtime up until summer.

The presence of grasses in an orchard (especially in dry soil) is not advisable for the first couple of years, as it may compete for nutrients and hydration.

Transplanting

The holes for transplanting must be dug in advance, taking care to distinguish the layers of soil when covering them again. Fruit trees should be planted on a slightly raised bed if possible. The neck and the rootstock should be just above the surface of the soil, except for species which, it is hoped, will quickly establish their own roots. There is a tendency to plant too deep. The grafting point should be about 10–15 cm (4–6 in) above the soil.

Use horn manure (500) or prepared horn manure (500P) at the time of transplanting; an essential act for the future health of the tree. After stirring it for an hour, spray the preparation on the soil and in all the holes; apply especially to the roots by adding to the root dip (a mix of clay, manure and 500P – see below, and chapter *Pastes and Root Dip*).

Transplant during a descending moon and avoid any unfavourable times shown in the planting calendar.

Pastes and root dip

It is important to put the roots into a root dip as soon as you receive the plants. See the chapter *Pastes and Root Dip* for more information.

Dusting

Maerl, clay, talc, basalt and wood ashes can be used as a powder, pure or in a blend to stimulate the tree defences and even prevent pest attacks (aphids, for example). Quantities used in dusting are one or two handfuls per tree (20–50 kg/ha, 18–45 lb/acre). Products containing calcium (maerl, ashes, dolomite, etc) improve the setting of the fruit.

Disease and pests

General fungal disease

Horsetail decoction, tree pastes and the use of horn silica (501) are basic measures to take against general fungal diseases. These can be supplemented by incinerating the diseased wood when the moon is in perigee and the sun is in the constellation of Aquarius (end of February, start of March). Grind the ash with a mortar and pestle for a whole hour and dry spray them on the diseased area of land (see chapter on *Pest and Weed Control*). You can also prepare a lactic fermentation with the diseased parts of the tree; stir daily checking progress (see also chapter on *Ferments*). Improve this fermentation by drying some of it, burning it, and adding the ash to the rest of the fermentation.

Canker

Scrape out and clean the affected part, and apply the pastes made from clay. Adding potassium permanganate (1%) or sodium silicate (1–2%) reinforces the activity. Dock *(Rumex obtusifolius)* tea or just rubbing the cankered zones with its leaves seem to give good results. Using slaked lime (calcium hydroxide) is just as effective as copper against apple tree cankers *(Nectria Galligena)* and possibly cankers of pear and plum trees.

The best approach is to avoid planting varieties that are sensitive to cankers.

Peach leaf curl

Several **preventative** measures can be taken.

- ❧ In spring and autumn apply tree pastes containing clay, manure, horsetail and basalt.
- ❧ Use lime sulphur distemper in winter or just before bud burst: 18–20 kg in 80 l of water (40–44 lb in 21 gal). It is preferable to make the distemper yourself.
- ❧ Harvest the diseased leaves and make a fermentation for the following year. Stir daily for a few minutes. When fermentation is complete, filter off the liquid and dry

out the residue (spagyric method); the solid residues are then burnt and the resulting ashes are dissolved into the ferment.

✤ Use clay (kaolin or bentonite) at 5–6%, adding an adhesive agent such as sodium silicate on the buds, before the dispersion of the spores. Saturate the buds. This seems to give good results.

✤ Suspending egg shells in a net hung from the tree itself produces results, though these vary from healing the plant completely to just a reduction in disease.

✤ Some gardeners suggest planting garlic at the foot of the trees, or placing copper pipes or old zinc roofing strips at the base of the tree.

Here are several **curative** methods:

✤ Sodium silicate at 0.5% applied as soon as leaves start to unravel from the bud.

✤ Propolis tincture at 0.5%.

✤ Lime sulphur distemper using 1 kg in 100 l of water; 500 l/ha are needed (4½ lb in 53 gal/acre).

As a last resort, use copper; Bordeaux mixture or Cuivrol at 1.5%, or copper oxychloride at the beginning stages of bud swelling. Apply very carefully. The leaves of the peach tree are particularly sensitive to copper salts.

Zinc trace minerals give positive results.

The amateur gardener should ideally plant resistant trees!

Citrus Sooty Mould

Instead of using mineral oils (white oils) you can use a spray with clay or sodium silicate as a base. This spray stops oxygen reaching the fungus. In certain cases, a starch application diluted in cold water has the same effect. Do not cover the entire tree as it still needs to breathe; only treat the diseased parts.

Alternatively use emulsified rapeseed oil (possibly using a mixer) in place of white oils.

Storage disease after harvest

In case of storage disease, look at crop management, starting with the choice of varieties, nitrogen fertiliser and irrigation techniques. Good lime nutrition on the trees is essential for good fruit storage.

Spraying horn silica (501) for three weeks up to one week prior to harvest helps storage. Horsetail decoction can be applied just before the harvest or even on the harvested fruit to reduce fungal disease during storage.

Spraying fermented whey sometime after harvest also gives good results.

Monilinia (brown rot)

Tree pastes and winter sprays, when the last leaves fall off the tree, and in spring when the buds starts to open, are basic measures, along with lime sulphur distemper. Horsetail decoction and horseradish tea also work (suggested by Jean Luc Petit). The use of a fermented tea from the leaves and fruit in question could produce interesting results.

It is important to carefully pick out shrivelled fruit and remove affected parts of the tree.

Powdery mildew

Use the same measures as for treating scab (see below). Milk or whey at a dose of 20 l/ha in 150 l of water (2 gal in 15 gal/acre) applied regularly (every 10 days) on vines gives good results. In the case of more widespread problems, use coarse salt at 2% (or potassium permanganate at 0.1%) dissolved in plenty of water (500–1000 l/ha, 50–100 gal/acre).

Phytophthora

To treat phytophthora, alternately spray horn manure (500 or 500P) and horsetail every two days. Saturate the entire plant, including the roots by applying a thick solution of tree paste to the trunk. Stop after 3 or 4 applications. Care for the tree normally and cut the grass surrounding the tree regularly; mulch should never touch the trunk. Make sure the tree has adequate light. Good drainage is vital.

Scab

Using copper salts, the most common product in organic agriculture, is limited by the Demeter standards to 15 kg/ha (13 lb/acre) of copper (metal) over a period of 5 years, with a maximum of 500 g/ha (7 oz/acre) in each application. This translates into 15 kg/ha (13 lb/acre) annually of Bordeaux mixture at 20%. Limit, therefore, your use of copper salts to treat scab.

Instead, plan for a programme of biodynamic sprays in between seasons: horsetail decoction and winter tree pastes. During the season, depending on the severity of the situation, use horn silica (501), horsetail, sodium silicate, and lime sulphur distemper, NAB at 1% or wettable sulphur at 0.6%, possibly adding sodium silicate (0.1%). Myco-Sin, as well as chive tea and dry spraying with clay, also give good results.

You can reduce the chance of reinfection by cutting the grass to accelerate the decomposition of fallen leaves, and burying the leaves superficially in winter. Spraying with horn manure (500 or 500P), barrel compost, fermented teas or even compost extracts has a similar effect. Avoid any application of copper in the off season as this can reduce the activities of the decomposing fungus and earthworms.

Amateur gardeners are advised to plant resistant varieties.

Animal pests

Before anything else, the immediate environment has to be rebalanced, and choosing preventative methods is a priority. Incinerations, the use of ash (dry sprays) and using D8 of the pest's ashes are specific practices that can also give promising results in certain cases (see chapter on *Pest and Weed Control*).

Weevils

In a balanced orchard, weevils are often a regulator that eliminates surplus fruit. In the case of serious problems, use Spinosad, an insecticide derived from soil bacteria, once or twice; pyrethrum extract can also be effective.

Red spider mites

Red spider mites indicate a nutritional deficiency or an imbalance linked to an overuse of phytosanitary products (sulphur, for example). In a case of infestation, use horn manure or nettle tea. Tyflodromus (a predator of the red spider) is often found in the flower borders of orchards, and the use of sulphur in large doses destroys these valuable beneficial insects. In case of a severe, uncontrollable attack, you can also try using paraffin oil or vegetable oils.

Voles

Curdled milk or fermented milk is a good repellent. Caper spurge (paper spurge, *Euphorbia lathyrus*) is also interesting because it produces latex. Elderberry leaf fermented tea has been tried and gives interesting results. Trapping these animals is essential (the topcat-trap is a high quality snap trap with a sensitive mechanical release mechanism). There are other preventative methods: not putting on mulch, working the soil, encouraging predators (weasels, martens, ferrets, skunks, foxes) in the environment, and using rootstocks that are unpalatable to rodents.

Making ash and dry spraying or D8 spraying can help, if for whatever reason the preventative methods have not produced results. Making roosts for birds of prey is another good control method.

Codling moth and moths in general

Fruit varieties with a hard skin are not palatable for the codling moth. Use lime sulphur solution and sodium silicate in the winter treatments, and apply clay-based tree pastes. Thin the fruit on the tree, picking up fallen fruit, and use horn silica (501) frequently. These are all important preventative measures.

Having poultry and bird feeders in the orchard are good measures that support diversity and help create a balance. Increasing the diversity of plants, flowers and shrub species in hedges and the local environment also helps. The use of essential oils to create an ambiance is an excellent step to take.

In the event of a heavy attack, products based on Bacillus thuringiensis (Bt), or granulose virus, can be applied at the moment the caterpillars hatch. As there are several hatchings during the year, use traps to find out when best to intervene. In difficult areas, professional growers apply Bt every 7 to 10 days and use pheromone attractants to disrupt mating.

In milder case, wormwood (*Artemisia absinthum*) and tansy (*Tanacetum vulgare*) teas give good results, as do clay-based sprays.

For heavily-infected orchards, the use of organic insecticides may be necessary, at least when first converting to biodynamics before a good balance has been achieved; ryanodine is recommended against moths, although it isn't authorised for use in some countries.

Sawfly

In case of emergency, Quassia extract gives good results.

Black and green aphids

A first measure is often to limit compost to reduce vitality.

Next, fill some flower pots with straw, hay or wood wool (excelsior), and turn them upside down. This offers some protection to colonies of earwigs, which will quickly develop and which are natural aphid predators. About one pot for every 2 or 3 trees is enough, and they can be moved depending on where the attacks are.

A preventative measure is to spray with rapeseed oil (1% oil

An earwig shelter

and 0.25% propolis tincture) before the buds come out; the oil prevents the eggs hatching. Repeat the treatment after a few days. This will also combat mealy bugs.

Nettle tea or ferment is possible but gives varying results.

If the infestation is a result of excessive plant vigour, the use of horn silica (501) may be effective. However, if the infestation is the result of plant weakness, use horn manure (500 or 500P), possibly stirred with a well steeped nettle tea.

Soft soap at 2% is a good measure at the beginning of infestation (before the leaves start to roll up). It can be supplemented with one or two tablespoons (10 ml) of rapeseed oil to 10 l liquid, and sprayed. Adding a wetting agent (pine terpene) increases the efficacy of the soft soap. Some ants cultivate aphid colonies; limiting the ants with glue rings produces interesting results in a household orchard. Do not apply glue directly to the bark as this can cause serious damage.

In a case of serious infestation, vegetable insecticides like pyrethrum or neem oil work well (but keep in mind the collateral damage they can cause).

Woolly apple aphid

Remove affected branches and burn them as a first measure. Treat branches containing aphids with a strong decoction of tansy or wormwood. Concentrated ferments of fern or nettle have been used with success. Several sprayings with calcined clay (such as Surround WP, Sokalciarbo and Argibio) in the autumn are effective. In case of serious problems, neem-based treatments are very effective. Use 2 l/ha (27 fl oz/acre) at the 'pink bud' stage (stage 5 of the BBCH scale), completed by 1 l/ha (14 fl oz/acre) after flowering.

Protection of birds

Birds are very beneficial for orchards and agriculture as a whole. For small areas (10th of a hectare, ¼ acre), install 2 to 3 nesting boxes, while for large plantations, 12 to 15 nesting boxes or more per hectare (5 or more per acre) are good. Various species of chickadees, red-tails, nuthatches, woodpeckers, swallows, bats, etc, should be protected as part of the environment.

Most damage done by birds can be prevented by installing permanent water baths and water sources. This is suitable in gardens and orchards, but can also be applied in grain fields for crows. The presence of nearby softwood is also a useful measure to reduce the damage done by some birds. As a last resort, burning the cadavers of crows (corvides) and then spreading the ashes as a dry spray on the areas affected has, in a number of cases, produced good results.

Water source for birds

Care after weather damage

Frost

After a sudden frost, if it is not too heavy, try using valerian extract (507); see chapters on *Plant Extracts, Herbal Teas and Decoctions* and *Climate Issues.*

Hail

After hail, apply valerian extract (507) as soon as possible. Dry spray with clay (or with a mix of clay and maerl) at 25–30 kg/ha (22–27 lb/acre) as a good supplement. After a hail storm, try a nettle tea. Depending on the weather after the storm, the plants will appreciate a spraying of horn silica (501) during the following days. See chapter on *Climate Issues.*

Compost in the orchard

The basics of biodynamic composting are essential, see chapter on *Biodynamic Compost.*

Improving the soil

Use about 5–12 t/ha (2–5 t/acre) compost per year, depending on the richness of the soil and the state of the trees. Some orchards producing 30–35 t of apples can almost do without compost if the biological activity of the soil and grass is good and the soil is rich and deep enough.

Is it better to add fresh compost or mature compost? According to Volkmar Lust, for small orchards, it is preferable to use mature compost (3–12 months old) as it resists disease better than fresh compost.

Organic fertilisers should not be used until the autumn.

Ideally, apply between the end of the harvest and mid November, but if necessary, during the winter up to the end of February. Spring composting can be very useful for trees that fruit early in the season, like apricot and cherry.

In soil where the organic material is blocked, bringing in young, organic matter in the spring can remedy these blockages and improve productivity.

Under young trees, spread the compost around the base, but avoid getting too close to the trunk. For older trees, spread inside the drip line. For fruit hedges, spread compost to 80 cm (30 in) on each side.

Vegetable composts are useful for orchards and vineyards, but it helps to add a bit of composted animal manure (horn manure or powder).

Poultry compost, in moderate quantities, also works well in orchards; there is a connection between the realm of birds and the realm of trees, as described by Rudolf Steiner in the Agriculture Course.

Apple pomace is easily made into good compost, incorporating a third to a half animal compost.

Compost supplements

In most young orchards, bringing in 500–800 kg/ha (450–700 lb/acre) of basalt per year is good at the beginning. Gradually reduce the amount depending on the state of the soil and plants. This measure is very useful in heavy, cold soil, as basalt stimulates warmth.

Generally in orchards the calcium level should be carefully monitored. The loss of this element is considerable, and in certain soils it must be added. In acid soil or where there is a calcium/magnesium deficiency, maerl or calcareous rock powders (300–500 kg/ha, 350–450 lb/acre) per year can be used in a reconversion period. Then if necessary, an annual dose of 100–300 kg/ha (120–350 lb/acre) can be incorporated into the compost.

For magnesium deficiency in calcareous soil with a higher pH level, kieserite can be used. Do not exceed 150–200 kg/ha (130–180 lb/acre).

Horn powder can be used at 500–800 kg/ha (450–700 lb/acre) annually, or for difficult varieties, up to 1000 kg/ha (900 lb/acre).

Bought organic compost is only a temporary solution. If you run out of local manure, you must think hard about the agricultural value and composition of bought composts. If using bought compost, to enliven and bring structure, do at least 3 passes of barrel preparation after spreading the compost. If the soil is sufficiently warm and damp, a single pass of prepared horn manure (500P) is enough. In orchards, supplementing this with foliar sprays is extremely important. Throughout the season, use various teas, vegetable extracts and fermented teas. Comfrey and valerian are good before the flowering stage. Nettle, comfrey, burdock, dandelion, camomile and yarrow are some of the teas that can be used during the growing season (see chapters on *Plant Extracts, Herbal Teas and Decoctions* and *Ferments*)

Care of the orchard through the year: a summary

The season starts in the autumn with spreading prepared biodynamic compost. Ideally, the time for this is shortly after harvest up to mid November. The quantity depends on the soil and plants, maximum 500–1000 kg for 1000 m² (2–4 t/acre).

Autumn spraying of 500P on the soil and under the trunk should be done from October to before the ground starts to freeze. If using barrel preparation, three passes are needed.

Spraying of clay, manure and horsetail solution, is very important when the leaves start to fall. The trunks can be painted with the remaining solution. If only one tree paste is applied per year, it is best to do it in the spring.

If significant fungal problems (monilia, scab, powdery mildew, etc.) have been observed during the year, before applying the tree paste, spray a lime sulphur solution (150–200 l solution at 10% per ha, 15–20 gal/acre), or apply the horsetail decoction at the beginning of spring.

It is important to spray of the whole crown of the trees again before the leaves bud. Horsetail is vital in this spraying. At this stage you can add a bit of sulphur, or use a lime sulphur solution. A sulphur spray or NAB solution can also be repeated just after flowering.

Horn manure (500) or prepared horn manure (500P) are used in the evening between the end of March and the beginning of May (100 g/ha stirred for one hour in 25–35 l rain water, 1½ oz/acre in about 3½ gal). Do this once or twice according to needs.

To regulate insects if there have been strong attacks in previous year, use soft soap preparation, but take care to avoid the flowering stage (one pass before, one pass after). Add some propolis tincture and several drops of essential oils (see chapter on *Products for Stimulating and Regulating Plant Health*). In case of a heavy invasion of caterpillars, Bacillus thuringiensis (Bt) based products are very useful.

Spray horn silica early in the morning on fruit already formed, but never during the flowering stage. Do this one to three times during the maturing of the fruit, two to three weeks before harvest, depending on weather. Some practitioners recommend doing these last sprays of horn silica in the evening, but our observations have shown that it is better to do this early in the morning.

During the season, maintain the growth and stimulate the immune system of the trees by regular use of teas, ferments, and decoctions, as needed. Nettle, horsetail, willow, comfrey, fern, etc. can be used on their own, or combined in a blend.

These treatments form a basis for a successful orchard, but they can be supplemented by other specific treatments, depending on circumstances. In a number of cases, if the environmental conditions are good, one can reduce these treatments, particularly reducing the use of sulphur.

Working the soil in the orchard

It is good to work the soil around the trees in the first couple of years (at least the soil below the crown). Thereafter usually the natural grass needs regular care by mowing or grazing.

Monitor the time of flowering and ensure that there is enough food for beneficial insecs (partial and alternating mowing can achieve this).

In dry spells or years, anticipate the situation by hoeing the soil under the trees or appropriate mulching.

If there is a heavy invasion of scab, a superficial disking of the leaves when they have all fallen (end of January to February) is a good controlling method. This should not be done too early because the winter cover is valuable in avoiding erosion and washout of mineral elements. Furthermore, the winter activity in the roots allow good microbial activity in the soil that encourages natural formation of humus.

Conclusion

For orchards in general, Volkmar Lust's recommendations are valuable, in particular his plan for reconversion. Some of his essential ecological methods are as follows:

- ꙮ Establish hedges and windbreaks.
- ꙮ Mulch anew every season.
- ꙮ Protect and feed birds, installing roosts for birds of prey.

- ☙ Introduce beehives (2–5 hives per orchard ha, 1–2/acre) and plant melliferous shrubs with staggered flowering times, beginning in early spring through to the end of autumn.
- ☙ Work with the rhythms of the seasons and the cosmos.

We would like to add the suggestion that a balanced orchard cannot really be achieved without animals. Even raising some poultry, or sheep, is an important step towards a balanced whole-farm organism.

18. Viticulture

The following suggestions are in addition to practising correct agricultural and organic principles. Stopping the use of weed-killers, and working the soil of the vine rows are important steps in the delicate transition from conventional agriculture. Plant care treatments authorised in organic viticulture are as reliable as those used conventionally, though in general they need to be applied more frequently. Effective and well maintained spraying equipment is essential for good plant health.

Careful plot monitoring is indispensable. Careful and regular observation of the weather will help develop an intuition to foresee what is coming. In difficult climatic conditions, it may even help to set up a well-sited weather station, to enable better control of plant treatments. This will enable further reduction of the use of copper.

The first year's work

The wine-producing year commences right after the harvest. Spray prepared horn manure (500P) in the evening, as soon as possible after the harvest. If this treatment cannot be done in time and the soil has already cooled, substitute this single pass of 500P with three passes of Maria Thun's barrel preparation, either on three successive evenings, or following the rhythm of trines at intervals of 9 or 10 days (see the *Maria Thun Biodynamic Calendar*).

After this, the prepared horn manure (500P) is applied in the spring in the evenings, between the end of March and the beginning of May, until the moment of bud burst or a little after. Earlier is better if the soil is sufficiently warmed and slightly humid. You can also add a few drops of valerian into the water at stirring, to combat climatic stress.

Vines showing growth problems (viruses, chlorosis, weak growth, etc.) will need a second pass of 500P. In serious instances, carry out treatment using a quadruple dose of horn manure (that is, 400 g/ha, 6 oz/acre), or pass again with 500 or 500P three times at weekly intervals (see the chapters on *Horn Manure* and *Horn Silica*).

Before flowering, make one or two passes of horn silica (501), stirred and sprayed very early in the morning, taking the weather into consideration. This can be done at the five-leaf stage, and when the tendrils 'pull at the vine'. Vines that are weak, aged, affected by virus or chlorosis should not be sprayed with silica until their growth has become well-established. If the growth does not become well-established, it will be necessary to wait until after flowering, or even until autumn.

A typical vine which has been sprayed with horn silica (501)

You could make one or two passes of 501 in early morning during summer or before picking, depending on the maturity and weather conditions.

Finally, to finish the cycle and start the next season, apply horn manure preparation (500 or 500P) as soon as possible after the grape harvest.

If needed, horn silica (501) can be used in the evenings at the end of October or beginning of November to accentuate the maturity of the wood and allow the reserves to migrate towards the roots. This practice is only useful if you have not been able to apply 501 during the season, or if there is not enough mature wood, or to stop vegetative growth starting after the season is over.

Suggested work for following years

The work in subsequent years will be similar to the first year but adjusted for the evolution of the soil, and changes in climate. The following measures can also be taken.

In the autumn after the harvest and before the end of November if needed, spread compost or organic manure. If biodynamic compost is not available, spray barrel preparation three times after spreading compost or, if soil conditions are right, spray prepared horn manure (500P) just once.

Tree paste with a mix of good quality manure tea, clay, horsetail decoction and whey can be used on the vine after the leaves fall. It is extraordinarily beneficial in the long term. There are a number of ways of making and applying paste: see the chapter on *Care of Fruit Trees*.

If the vines suffer from any kind of wood disease, use a thick paste or Cade oil (juniper). Another possibility is to make a paste like the one used in the autumn after pruning or at the end of winter before bud burst (see chapter *Pastes and Root Dip*).

If a vine is sensitive to mildew, spray horsetail decoction on the ground before bud burst (before Easter) at a dose of 100 g/ha to 30–150 l of water (1½ oz/acre to 3–16 gal).

On vines showing difficulty in growth (virus disease, chlorosis, weak growth, etc.), spray the vegetation with reinforced horn manure (400 g/ha of 500 or 500P stirred in 35 l water, 6 oz/acre in 4 gal) with a strong nettle tea added.

Where vines have had a very heavy dose of fertilisers and pesticides and have not reacted to any biodynamic preparations within two years, spraying a homeopathic dilution of Thuja D30 can help to release the blockage (see chapter *Products for Stimulating and Regulating Plant Health*).

You can judge the positive effects of biodynamic preparations by observing changes in the soil structure and colour, and its ability to drain. Also observe the accompanying flora and plants: their ability to stand erect, the colouration of the leaves into a more luminous green. And finally, note the development of resistance to disease, the improvement of organoleptic qualities (aspects of food experienced by our senses), and the quality of the harvest. All are good indicators of a healthy vineyard.

Mildew and minimising copper doses

Each vineyard and every area of land has its own specific climate, exposure, varieties, rootstocks, growth patterns, and in particular, levels of excessive vine growth. It is therefore impossible to give precise indications, but a few general considerations can be useful.

As we've previously noted, Demeter standards limit the use of copper to 15 kg/ha (13 lb/acre) of copper (metal) over a period of 5 years (that is an annual average of 3 kg/ha, 4½ lb/acre) with a maximum of 500 g/ha (7 oz/acre) in each application.

As well as basic agricultural measures (like limiting excessive vigour) and spraying horn manure (500 or 500P) and horn silica (501), minimising copper doses requires spraying horsetail decoction in the spring, as well as adding horsetail to all plant health treatments, whether teas or blends, mixed with copper and sulphur. Use horsetail, nettle, willow, elderberry, etc. in succession or in a blend.

Horsetail decoction applied at moon's perigee, especially around full moon, helps to strengthen the resistance of vines. Using horsetail decoction with Myco-Sin or copper in this case is highly recommended but don't spray during flowering period (spray either before or after). This also applies to powdery mildew.

It is better to start early with small doses of copper instead of allowing the disease to take over, and having to sort out the mildew problems throughout the season. Small doses of copper can be repeated frequently, progressively increasing the dose every one to two weeks, depending on the local context, climatic conditions, growth and variety of the grape. Renew after 20 mm (¾ in) of rain, and do not allow more than three or four young leaves to be unprotected. In almost all cases, if there is reasonably good health and low pressure from mildew, you can start with doses of 100–200 g/ha (1½ –3 oz/acre) of copper (metal) and maintain or increase according to need. The manufacturer's recommended dose can, in a biodynamic context, be divided by 5, or even 10. Only high yielding vines growing in the northern wine regions, or vineyards in humid maritime conditions, may have difficulty staying within the Demeter standards if there are many difficult years in succession (see Copper salts in chapter *Products for Stimulating and Regulating Plant Health*).

In case of serious mildew on grape clusters, a powder made from deployed copper carbonate at 30 kg/ha (27 lb/acre) is quite effective. However, it is a high dose of copper in one single treatment, exceeding Demeter standards for a single application (30 kg is equivalent to 3.75 kg copper metal). It is better to use Matthias Wolff's Fire remedy.

At the end of the season use hand-held shears to cut out young leaves affected by mosaic mildew and the contaminated suckers in between. This will allow the situation to be tackled without using additional copper.

In a number of cases, it is possible to significantly reduce the amount of copper. All the biodynamic practices need to be carried out well, the landscape must be diverse, its animals need to provide closed-system compost, the vines must be in a balanced state, and the yields relatively low. It is possible to use plant extracts, for instance Myco-Sin, various teas, decoctions or essential oils. However, this requires a lot of experience and is therefore not recommended when first introducing biodynamic practices.

Powdery mildew and minimising sulphur doses

In certain cases, one may want to try to reduce the sulphur dosage used against powdery mildew, either because of recurring problems of reduction in the cellar, or because of problems of acidification in soil that is already too acidic.

For this, milk or whey is effective. Large quantities are

needed, 25–50 l/ha (2–5 gal/acre). Acid whey (obtained while making acid cheeses, like cottage cheese) is most often used, although sweet whey works as well. In pre-flowering anti-mildew treatments, whey gives good results when alternated with sulphur. Using yarrow tea at 10–50 g/ha in 35–150 l water (4–6 ml/acre), on its own or together with other classic treatments of copper and sulphur, can minimise the dose of the latter.

Fennel essential oil can be helpful, especially towards the end of the season. It works by limiting the amount of sulphur needed for varieties sensitive to reduction.

The use of sodium bicarbonate or potassium bicarbonate is also something to try, particularly in acidic soils. Use sprays at a dose of 6–8 kg/ha (5–7 lb/acre). However, this does not seem to produce good results on Mediterranean vineyards.

Products using fenugreek like Stifénia used before the flowering stage appear to give good results; use the recommended doses.

Finally, lime sulphur solution should not be ignored, for it allows for a reduction in sulphur doses. Calcium polysulphate is also effective, even in cold temperatures.

The addition of terpenes gives the sulphur a better adherence and makes it longer lasting.

Preventing powdery mildew

On land where mildew is persistent, burning contaminated pruned branches and spreading these ashes has proved to be a preventative measure. Incineration is done when the moon is in perigee and the sun is in the constellation of Aquarius (February/March), and if the ashes are spread as quickly as possible. Sprinkle the ashes in small quantities, like pepper on soup. To make the spreading easier, mix the ashes with some sand, basalt or maerl, whatever is available and depending on the needs of the land.

In the case of serious contamination from the previous year, use a lime sulphur solution after the leaves fall and before winter tree pastes are applied.

Tree pastes are also a good controlling method: in this case, add a strong decoction of horsetail tea, or sodium silicate at 2% (or even both together), to a manure tea and clay solution.

In regions where mildew is a significant problem, and for sensitive types (Carignan, for example), start the sprays and powders as early as possible, when one or two leaves start to spread. Dusting sulphur powder with a pierced box allows use of only 10–12 kg/ha (9–11 lb/acre) of sulphur. In some cases, spraying liquid sulphur at the beginning of the vegetative stage will be a determining factor for the rest of the season. Spraying between the stage of the buds showing a green point, to the bursting of the buds, with sulphur and horsetail tea, can be an excellent measure against powdery mildew and even against dead arm.

Dry spraying around flowering is both a good preventative measure, and for situations of high risk; however, you must protect the flower by only spraying immediately before or after flowering. For this use 15 kg/ha (13 lb/acre) of sulphur mixed with the same amount of clay or talc.

Curing powdery mildew

A number of procedures have been used, with varying efficacy. First, use a powder at high dose. Normally, it is not recommended to exceed 15 kg/ha (13 lb/acre) of sulphur, but as it is not possible to spread such a small quantity of powder evenly, it is best to mix this sulphur with equal parts of talc, bentonite clay or maerl.

In cases of strong contamination, table salt at 2–3 kg/ha greatly diluted with 500–600 l of water (2–2½ lb/acre in 55–65 gal) has shown results. Add 10–12 kg/ha (9–11 lb/acre) of wettable sulphur in this saline solution.

Potassium permanganate has often been used at a dose of 0.125-0.15% in 400–800 l/ha (45–85 gal/acre) of water. With few exceptions, it is not authorised nowadays. If this method is used, it is absolutely essential to spray with wettable sulphur at 10–12 kg/ha (9–11 lb/acre) in the following hours (because permanganate weakens the protective cuticle of the grapes).

In some cases, horn silica (501) stirred for an hour has produced positive results. A large quantity of water is needed: enough to drench the foliage. Using 0.5% hydrogen peroxide in 600 l/ha water with 10–12 kg (9–11 lb/acre) wettable sulphur has given variable results. Whey or milk sprayed directly on the clusters has given positive results. Removing leaves around the clusters to aerate and give more light is an essential measure, but take care that the clusters are not sunburned.

Various diseases and pests

Scale insects

Preventative measures against scale insects are important: good soil development, 500 or 500P, compost, and autumn and spring pastes to clean the vine. In acute cases, soft soap can be used during the growing season.

Here is an example from south Burgundy: dilute 8½ l liquid soft soap, 22 l 50% alcohol and 12½ l colza oil in 400 l water; this is enough to spray one hectare (1 gal soft soap, 3 gal alcohol, 1½ l colza oil, 40 gal water per acre). Be sure to mix well. Spray a second time, about one hour after the first spraying; the first pass causes them to lift their waxy wing scales, and it's the second which truly reaches them.

In addition, in serious cases, lime sulphur can be used at the end of the winter.

Cluster worms

Pheromone traps diffuse synthetic substances into the environment. This raises a number of questions about long term effects on human health as well as control of insects in the vineyard. A number of vineyards report reduced efficacy against grapevine moth *(Lobesia botrana)* or cochylis or vine moth *(Eupoecilia ambiguella)*, and sometimes an increase in numbers of pests that are more difficult to manage, like spotted cutworms *(Xestia adela)* and willow beauty moths *(Peribatodes rhomboidaria)*.

Bacillus thuringiensis (Bt) can be used once or twice per generation, depending on pressure. Apply when hatching begins, or even before (when the heads are black and visible in the egg). This can be completed by burning some moths and spreading their ashes. According to Maria Thun, for the grapevine moth, it is best when moon and sun (and if possible Venus) are in Gemini, and for cochylis, a nocturnal species, when moon and sun (and ideally Mercury) are in Aries.

Spinosad is approved for organic use, limited to two uses per year on the same infestation. However it is less environmentally-friendly than Bt.

Bees can help control grapevine moths, and bats are important predators of nocturnal or semi-nocturnal moths; bats can be encouraged by introducing bird nesting boxes for use as temporary lodges.

Black rot

Black rot, particularly bad in some regions, is on the increase. Copper can be used when the grapes are just visible and four to six leaves appear, but preventative measures are more important. Systematically pick out and burn the contaminated leaves, eliminate contaminated clusters and dried fruit that remain on the vine. Burn the contaminated wood when the moon is in perigee, and spread the ashes on affected areas. Sweet chestnut bark tea and ash bark tea have given good results in some areas; trace elements such as zinc, boron and manganese could also be explored.

Bugs that eat buds (cutworms and willow beauty moths)

Maintaining a minimum of grass allows caterpillars to find enough food without being tempted to eat young shoots. Spray horn manure preparation (500 or 500P) when the caterpillars start doing damage, to accelerate the budding process and bring a regulating effect through a substance of animal origin.

Preventing these problems needs a look at the presence of animals that can create a balancing effect around and on the vines. Birdhouses and beehives nearby play a positive role. Poultry can scratch and eat the caterpillars, which can be very helpful. Some wine growers have made moveable chicken coops to bring chickens to the vine during the pruning work. One Chablis wine grower even constructed a wind-tunnel machine to collect moths.

Traps based on Bacillus thuringiensis (Bt) or pyrethrum mixed with wheat bran have not been very effective. Bt does not work well in the spring as it is still too cold. If one is faced with serious insect pressure, try using Bt in the autumn.

Wood disease, esca, black dead arm, eutypa dieback etc.

Questions need to be asked about the practices in many nurseries, in particular omega grafting, disinfecting wood with Cryptonol, the use of hormones and general over-fertilisation.

Good hygiene while pruning is essential: disinfecting the pruning shears, taking out and burning wood that is more than two years old, painting pruning wounds. Eutypa, for example, can be controlled by late pruning and good hygiene.

Esca, a wood disease

Cutting back the stock as soon as symptoms appear gives a temporary remission. Drilling holes does not seem to give great results. However, making a slit in the stock with a wedge and hammer, and inserting a pebble to keep it open, or cutting a large mortice (15–20 cm, 6–8 in) at the base, brings about internal aeration and introduces light, which prevents the development of fungus. Use St John's wort oil, or substances containing salicylic acid like elderberry, willow and meadowsweet, to complete the process. Daniel Noël has suggested the use of a paste of eucalyptus essential oil with hydrogen peroxide used after pruning.

Grapevine yellows and degenerative disease

In some countries the use of pyrethrum is rendered compulsory to stop grapevine yellows. However, it is not the best solution as pyrethrum is a non-selective insecticide and so destroys efforts to create local biodiversity. A better solution to these problems is to stimulate a genuine regeneration of the vine through balanced ecological measures and keeping to biodynamic principles. Clay (kaolin) in a liquid spray appears to give good results in controlling leaf hoppers, as do essential oils (garlic, fennel, lavender).

Topping and trimming

Letting the vine regulate itself is ideal. To do this, maintenance of the apex (final bud) must be left as late as possible in the season. Cutting the apex too early stimulates formation of secondary branches and sets off growth activity again, during a time when fruiting should predominate. The first topping should not be done until the apex begins to bow.

Regulating vigour by manuring and pruning are fundamental principles to keep vines balanced.

Reduce the number of trimmings and toppings, by doing the former as late as possible, after Saint John's Tide (June 24). The descending moon is favourable, and avoid working on days around nodes, perigee, and avoid leaf days (as indicated in the planting calendar) which stimulate the vegetative process.

Spraying unstirred valerian at the time of trimming helps to relieve the stress of the cut, heals the scarring, and reduces the formation of secondary branches. Use 5 ml in 50 l rainwater (1 tsp in 13 gal).

It is sometimes possible to reduce the amount of pruning by training and tying shoots to the trellis wires. Repetition of this practice helps regulate vigour and improves the quality of the harvest.

Pruning

Prune during a descending moon, ideally from February to the end of March, to help regeneration and vigour. It's best if the plants are already well balanced. Prune vulnerable vines, where possible, on fire-warmth and air-light days, such as Leo, Libra and Gemini.

Working at perigee can encourage vigour but don't do it too often since it can lead to uncontrollable growth and fungal disease. To reduce vigour, prune during an ascending moon in February and March.

Wood from the year can be reused by shredding the clippings or taking them out of the area to compost, depending on the situation and the capacity of the soil to assimilate. Take out and eliminate wood that is more than two years old, and if necessary burn heavily mildew-contaminated shoots.

Tree pastes and other coatings

See chapter *Pastes and Root Dip.*

Hail

After hail, try to intervene as soon as possible. Tractor spraying may be difficult if the soil is too soft and wet.

Valerian is valuable in this situation for both the grape and the vineyard. Make a solution using 5 ml in 30–35 l of water (1 tsp in 8–9 gal). Stir for 10 to 20 minutes, and spray with a backpack sprayer or atomiser in the hours following a hailstorm. This helps to relieve the stress of the impact of hail pellets and the resulting cooling. If valerian is not available, use nettle tea (possibly add a few drops of arnica tincture).

Dry spraying with clay or Luzenac talc within 24 to 48 hours also has a positive effect.

In the days or weeks that follow, consider one or two sprays of horn silica if weather conditions allow. Adding valerian (5 ml/ha, ⅓ tsp/acre) to classic treatment mixes (copper, sulphur, various teas) is a useful option if mildew or powdery mildew continue to be a significant problem.

Using copper immediately after a hail storm late in the season has not given good results. If there is a serious mildew problem, use a decoction of horsetail.

For more on weather issues, see chapter *Climate Issues*.

Working the soil and grass

Aim to leave some grassy ground cover between the vines through the summer, and preferably in autumn and winter, depending on climate, soil and grape variety. Ideally, natural, spontaneous grass growth is sufficient, encouraged where necessary. In springtime, a superficial, regular working over is needed to maintain effective control of early grassy growth.

Vine rows with grass (rye and white clover)

Ridging the soil at the end of autumn does not allow the plants, and therefore the soil, to develop well during the winter. Generally, it is better to work the soil lightly at the end of winter or the very beginning of spring, if necessary making only small ridges and flattening them again soon afterwards if there is a risk of too much grassy growth. An exception to this is heavy clay soils during a damp spring which might be unworkable otherwise; in autumn, make small ridges on part of the soil, alternating the following year. (Do not start any work on the soil until it is either dry enough or you are sure of at least a few sunny days to allow the freshly lifted vegetation to dry.)

Delaying work on the soil until spring often has a regulating effect on the vine, diminishing vigour and thereby contributing to growth stopping at the end of June.

In most cases, it will be necessary to work under the vine row during spring and in the beginning of summer. In certain cases, growing grass under the row is possible, but it will need regular cutting by scythe or strimmer. (If using a strimmer, make sure that the plastic cutting wire cannot damage the base of the stock.)

Replanting vines

The optimal time between plantings is seven years (although this may not always be possible). When replanting, the aim is to re-enliven the soil so that the left over root debris will decompose. Worms play a very important role as their secretions inhibit the development of nematodes, carriers of viral disease.

After harvesting, rip out and remove as many roots as possible by passing a chisel plough or cultivator. In autumn, spread good quality compost which has received the biodynamic preparations (5–10 t/ha, 24 t/acre), has aged more than six months, and is colloidal in nature. Turn it in very superficially. In calcareous soil, bring in some basalt at 500–1000 kg/ha (450–900 lb/acre) depending on need, and possibly kieserite (200–400 kg/ha, 180–350 lb/acre) if there is a deficiency in magnesium. In soil that is deficient in calcium and magnesium, use 250–800 kg/ha (220–700 lb/acre) maerl, depending on needs.

As soon as possible, sow a green cover crop, as diverse as possible (Matthias Wolff's mix, for example), having sprayed the soil with prepared horn manure (500P) either the day before or the same day. Doing this on a root day during descending moon can optimise the work, but the essential thing is that it is done early and under good agronomic conditions.

Sample autumn mix for a green cover crop

The following metric quantities are per hectare (US measures are per acre).

Seed of 100 kg (90 lb) of six grain crops (rye, oats, triticale, spelt, wheat, barley, in order of priority), vetch 50 kg (44 lb), winter peas 50 kg (44 lb), field beans 30 kg (27 lb), mustard 5 kg (4 lb), rapeseed 2 kg (2 lb). Then spray 500P on the soil.

If growth is rapid, and there is sufficient water, apply horn silica (501) on a leaf day in spring. Cut and turn in with a field shredder as soon as the flowering stage has started, at the end of April or beginning of May and during a descending moon, if possible. Immediately afterwards pass with the spring tine cultivator, to prepare the soil for reseeding and spray reinforced horn manure or 500P in the evening. (If you have to bring in slow-acting fertiliser (calcium, magnesian rocks) or more compost because of a soil deficiency, you can do so at this time.) Reseeding should be done as soon as possible, when the soil is still quite moist, with the mix suggested below.

If greater diversity is needed, or to compensate for a soil deficiency, add seeds from appropriate medicinal plants to these mixes.

Samples of mixes for a spring green cover crop

This type of green cover is ploughed under at the end of June or beginning of July, before the mustard starts to seed. Spray the 500P on the ground immediately after cutting. Leave the soil as it is, without sowing anything afterwards. The new growth will develop in the autumn. If the crop is too high, remove it otherwise the overabundance of mulch will impede new growth. You can either make hay or silage with the crop, or compost it. Use either the Matthias Wolff mix (see chapter *Green Manure*) or the following mix:

	Seed Mix A kg/ha (lb/acre)	Seed Mix B kg/ha (lb/acre)
Six grains	100 (90)	100 (90)
Vetch	30 (27)	25 (22)
Peas	30 (27)	25 (22)
Field Beans	30 (27)	25 (22)
White clover		1 (1)
Red clover		1 (1)
Untreated sunflowers		2 (1)
Mustard	5 (4½)	3 (3)
Buckwheat	10 (9)	
Phacelia	5 (4½)	
Rapeseed		2 (2)

If climate or irrigation allow, the ideal is a new sowing after working the soil with a tine cultivator. In that case, introduce summer plants with a short cycle like phacelia, buckwheat, crimson clover or Alexandria clover, or more complex mixes. Barley helps to deter nematodes.

Ideally continue this type of green cover cropping for

several years, each year increasing the diversity. Try harvesting grain (oats to remove copper, if necessary), followed by a green cover crop in the autumn. The optimum is an interval of seven years between replantings. However, economic constraints often do not allow for such a period of non-production. An intensification of biodynamic methods can support faster soil development.

In the year before replanting, turn in the vegetation using a tine cultivator sufficiently early at the end of autumn or beginning of winter to ensure a good healthy soil, without fresh organic matter, for the planting of vines in spring.

Other mixes

Another option is to seed alfalfa. Even better is a widely diverse meadow, with at least 5 different grasses: English or Italian ryegrass, tall fescue, meadow grass, timothy, cocksfoot in small quantities. Alternatively, seed 5 to 8 species of legumes: different clovers (white, dwarf, red, etc), alfalfa, melilot, sainfoin, trefoil, sweet clover. Harvest the crop and make hay over several years, and if necessary add compost at the end of summer or beginning of autumn. When growing meadow crops, use prepared horn manure (500P) and horn silica (501).

Turning in the meadow is done using several passes with a tine cultivator in the autumn preceding replanting. At this time sow a green cover crop, like phacelia that will easily be taken by winterkill and leave a good structure of soil depth. The preparation of the soil before replanting is then limited to a pass with the cultivator without disturbing the deeper soil.

Transplanting

Avoid deep cultivation of the soil either by ploughing or trenching. A light working with a tine cultivator is more efficient, and does not turn over the structure and life of the soil. During the preceding year, biodynamic preparations should be applied intensively, with a least one pass of 500P in the spring followed by one pass of horn silica (501) and two passes of 500P between the end of August and the end of October. Transplanting should be done during descending moon, if possible on a root day, but in all cases avoiding unfavourable times (see planting calendar).

Using root dip with all young plants and preparing the soil tbe year before give superb results in the development of young vines (see chapter *Pastes and Root Dip*). Transplanting should immediately be followed by spraying 500P and painting the plants with a mix of manure, clay, basalt and 500P. 500 or 500P can also be repeated in the weeks following transplanting, especially in dry conditions.

Choice of vine stock

Choose vine stock that have not had hormones used on them, either in grafting or re-rooting, and coming from reliable organic or biodynamic nurseries. Avoid plants in pots that have aged too long, showing roots turning upwards. Even if these roots are pruned they will have tendency to grow in this injurious way after transplanting.

Spray the young plants with horn silica (501) when they have a good showing of leaves (at least 5 or 6), preferably on a root day. This is important for the rooting and future development of the vines. If growth is weak because of climate or other conditions, don't spray 501 until the autumn or following year.

In case of drought or weakness, spray horn manure (500 or 500P) on the leaves. Nettle tea, comfrey extract, yarrow or camomile tea are also good.

Vinification

Winemaking must respect the biodynamic vinification standards. These standards permit few additions, and prefer natural yeasts. Wine is a living thing, enhanced by using biodynamic preparations, and it is important to preserve that life as much as possible in the wine cellar.

Grapes treated with biodynamic preparations have higher acidity (lower pH, and more stable) with less malic acid than conventional vines. The phenols are well rounded with ripe seeds, rich colour and good tannins. The grape juice is fresh and well balanced, with a long aftertaste – aromatic and vibrant. If grape growing and vinification respect the fruit, those qualities will be found in the finished wine.

If appropriate environmental measures have been taken and complementary fauna are present, and if the use of copper towards the end of the season is limited, wild indigenous (ambient) yeast, full of vitality, will be present on the grapes. If biodynamic preparations have been used correctly, the minerality of the terroir will be reflected in the complexity of the wine. This often results in a well-balanced, pure wine with excellent drinkability. Sensitive crysallisation images are helpful for revealing the characteristics of different wines. (For more on this, see *Sensitive Crystallization: Visualizing the Qualities of Wine* by Christian Marcel.)

Vines treated biodynamically have a stronger link to their terroir and are more resistant to climatic variability (for example, staying fresh and tender during dry years, and plump during years where maturation is difficult).

A well-grown grape can mean simpler work in the wine cellar, although it still, of course, requires great care, especially in the management of temperature and working rhythms. Wines produced biodynamically are more resistant to oxidisation and can be protected with less sulphur (yarrow tea applied during the growing season can also help with this).

Working with cosmic rhythms

A waxing and ascending Moon is favourable for yeast reproduction, and if stuck fermentation arises, it can most easily overcome at the full Moon. In many cases of stuck fermentation, spraying a few drops of valerian diluted in warm water and misting around the area can help revive fermentation. Some winemakers have had good results adding one drop of valerian per bottle.

Many winemakers have found that fruit or flower days (see the *Maria Thun Biodynamic Calendar*) are best for clarification, racking and bottling; they avoid watery days such leaf days, perigee and the full Moon, as well as lunar and planetary nodes. Days of atmospheric high pressure are better for bottling.

Ascending daily and monthly rhythms (morning, lunar spring) favour oxidisation, while descending rhythms (afternoon and evening) and the descending Moon favour reductive processes. The descending Moon also favours wines with aging potential; the ascending Moon favours young wines for early drinking.

It's possible to experiment with working in the cellar on different fruit days (Leo, Sagittarius and Aries); it could be very beneficial to understand more about the effects of different constellations on the qualities of wine (see also the biodynamic wine calendar, *When Wine Tastes Best*).

Racking and bottling on root days may add minerality, although differences seem to even out over time. Some winemakers also 'stir up' the lees (*battonage*) on particular days, noting the influence of the corresponding planet (eg. Monday Moon, Saturday Saturn, etc.). This provides a range of different times according to the needs of each cuvée and the style of wine required.

There is much more to understand about how cosmic rhythms work in the wine cellar.

Viticulture today

One of the major problems today is the over-specialisation of land used exclusively for viticulture. In some cases, this monoculture is confined to one grape variety, sometimes even to one particular clone. Research into floral and animal diversity shows that we need to start re-balancing these landscapes, creating integrated ecosystems.

In striving towards the ideal whole-farm organism, we need to look at local self-sufficiency of manure and compost, companion planting and floral borders, and the presence of animals (poultry, horses, sheep, etc.). Working to enhance the landscape (thickets, hedges, dry stone walls, etc.) can complete this circle.

Another issue is that a great deal of research still needs to be done to find types of vine resistant to disease, that adapt to the terroir and make good quality wines. For example, should

Poultry among vines

growing seedlings be left to nurseries or should it be integrated into the terroir? The early development of a plant – whether an annual plant, a tree or a vine – is a determining factor for its future health and longevity. Situations leading to atrophy (rare) and hypertrophy (more frequent), for example, have a detrimental effect on the future of the plant.

Developing local biodynamic nurseries is therefore of prime importance. Plants should be grown using biodynamic principles right from the start: a respect for cosmic and earthly rhythms and the use of biodynamic preparations on vines to be used for grafting. Hormones cannot be used in grafting, rooting and pruning; instead, the use of horn manure (500 or 500P) and clay paste while grafting should be encouraged. Disinfection should be carried out by using horsetail and propolis extract, or by preparations based on bacillus subtilis. The soil of the nurseries should receive biodynamic compost, and synthetic pesticides must be totally excluded. Such nurseries would ideally work with *selection massales* (a process whereby cuttings from many carefully-selected plants are used for grafting), and with rootstocks cultivated under good conditions without being forced with fertilisers (even organic ones).

Less intrusive whip grafting should be preferred to omega grafting because the latter favours necrosis and wood disease; cleft grafting is also a possibility for transplanted rootstock (or better, T-bud grafting). Lateral grafts best respect the physiology of the vine and avoid disease.

Regulations concerning vineyards demonstrate an increasing lack of understanding for the biosphere and nature of life, creating genuine obstacles towards regeneration of this field. In particular, exclusive use of shoot-tip micropropagation or microcuttings, or meristem-tip cultures that try to annihilate viral diseases, artificially remove the plant from the sphere of life.

We must research in other directions: into the effects of

Animals near vines

herbaceous grafting and other techniques less traumatic for the plant and less harmful for the future. Some promising research, directly related to biodynamic and anthroposophic concepts, explores the possibility of regeneration by taking a step back to look at the overall vegetative and growth forces, while preserving the character of the type of vine.

Wine and the grower

If the suggestions above are followed, disease problems will steadily diminish, and the subsequent work in the cellar will be much simpler. A perfect grape is the precondition of an authentic wine. It starts with an almost religious respect for the soil and the endeavour to harmonise the plant with its terrestrial environment and the surrounding cosmos.

This requires, as we've already seen, developing the diversity and complexity of the farm, treating it as a whole living organism, working with the rhythms of sun, moon and planets, as well as using the biodynamic preparations. On a personal level for the farmer or wine grower, it requires opening ourselves to the four elements of earth, water, air and warmth. This will awaken the artist within us, and bring about an inner development through

Light ploughing between the rows with a horse

observation, contemplation and meditation. Then we shall be able to re-establish the vital links between nature, humanity and the universe.

Conclusion

In my work as biodynamic advisor I have visited different biodynamic farms throughout France and beyond, and have witnessed approaches which range from those valuing almost total self-sufficiency, to those pursuing instant results to a degree that is unreasonable. The trends mentioned above radically alter the initial impulse. The starting impulse should be to put oneself in a position of responsibility vis-à-vis the evolution of the Earth and of human beings. It's a question of learning to understand nature and its laws, then putting them into practice to bring about prosperous and sustainable agriculture, to improve human nutrition and the health of the Earth.

To bring about increased biodynamic agriculture, we must

start from a base of good, professional agricultural practice and an understanding of organic methods. We must build on these to help biodynamics come to fruition and be taken up by a larger number of farmers and gardeners.

The urgency of environmental demise, and the significant role agriculture plays in the modifying ecosystems and climate, mean that biodynamic agriculture is of growing importance. It truly answers some of the most pressing questions of our time.

Its philosophical basis also brings an understanding of the links between human beings and nature in all their complexity, and begins to restore plants, animal and human beings in their right relationship with the cosmos and the earth.

The ideal is that farms move towards being whole, self-sustaining individualised organisms. Every country and more importantly every region needs to ensure local food sovereignty and security. Autonomous food supply and seed saving initiatives are vital for developing countries, but this vision is also relevant for developed countries. In developing countries, the diversity and autonomy of most farms often increases soil fertility and can help create a harmonious landscape which can also nourish thousands. Diversified farms which also breed domestic animals need to be encouraged.

This agricultural approach responds to the growing environmental concerns of our time. The earth, seen as a living organism, seems to be showing a phase of ageing, of mineralisation, of hardening and of salinisation, becoming sterile in some parts. Human activity – through the use of mineral fertilisers and pesticides, a lack of organic restoration, and acquisition of heavy agricultural plant (machines) – is only accelerating this process. Using biodynamic preparations can dissolve mineral formations, making the soil more pliant and vital again (thus saving energy), and better able to withstand erosion. Trials by the Research Institute for Organic Agriculture in Switzerland (FiBL) and by Dr Edwin Scheller have shown that biodynamically-treated soil forms more humus and microbial matter than with other agriculture practices. This is a positive contribution to reducing carbon dioxide, necessary to combat global warming.

Biodynamic agriculture has been proven to maintain and improve soil fertility. It brings about structure essential to life: these soils retain water and air better; they breathe more easily and are more alive. The resultant productivity, and nutritional balance and quality, can be seen immediately.

Biodynamic food has better nutritional qualities, for example, enhanced levels of vitamins and minerals. The higher levels of silica in biodynamic plants are consistent and result in better storing qualities. This nutrition benefits the human immune system and general wellbeing, as shown in trials by Dr Karin Huber and at the Institute for Biodynamic Research in Darmstadt, Germany.

Other biodynamic products are recognised throughout the world under the collective international trademark of Demeter, which controls and certifies biodynamic practices. Nutritionists, therapists, pharmacists, producers of cosmetics, textiles, as well as consumers, appreciate the quality of these products.

Responsibilities are shared. Biodynamic producers carry the burden of implementing biodynamic methods and developing a whole-farm organism. Consumers also have a responsibility, however. They must support this work, being prepared to pay a fair price for products and recognising the value and quality of the work. Positive, balanced relationships between producers, processors, sellers and consumers must be established, to mutual benefit.

Our world is facing unprecedented change. Faced with difficult political, economic and social questions, the principles of biodynamics can offer some important answers.

Bibliography

Cloos, Walther, *The Living Earth*, Lanthorn Press.

Colquhoun, Margaret and Axel Ewald, *New Eyes for Plants*, Hawthorn Press.

Conford, Philip, *The Origins of the Organic Movement*, Floris Books.

—, *The Development of the Organic Network*, Floris Books.

Karlsson, Britt and Per, *Biodynamic Organic and Natural Winemaking*, Floris Books.

Klett, Manfred, *Principles of Biodynamic Spray and Compost Preparations*, Floris Books.

Koepf, H.H. *The Biodynamic Farm*, SteinerBooks, USA

—, *Koepf's Practical Biodynamics*, Floris Books.

Kranich, Ernst Michael, *Planetary Influences upon Plants*, Biodynamic Farming & Gardening Association, USA.

Lepetit de la Bigne, Antoine, *What's So Special About Biodynamic Wine?*, Floris Books.

Marcel, Christian, *Sensitive Crystallization*, Floris Books.

Osthaus, Karl-Ernst, *The Biodynamic Farm*, Floris Books.

Pfeiffer, Ehrenfried

—, *The Earth's Face*, Lanthorn Press.

—, *Pfeiffer's Introduction to Biodynamics*, Floris Books.

—, *Soil Fertility, Renewal and Preservation*, Lanthorn Press.

—, *Weeds and What They Tell Us*, Floris Books.

Pfeiffer, Ehrenfried and Michael Maltas, *The Biodynamic Orchard Book*, Floris Books.

Philbrick, John and Helen, *Gardening for Health and Nutrition*, Garber, USA.

Podolinsky, Alex, *Bio-dynamic Agriculture: introductory lectures*, Gavemer Foundation 1985.

—, *Biodynamics: Agriculture of the Future*, Bio-Dynamic Agricultural Association of Australia 2000.

Sattler, F. & E. von Wistinghausen, *Biodynamic Farming Practice*, Biodynamic Agricultural Association, UK.

Sattler, F. & E. von Wistinghausen, *Growing Biodynamic Crops,* Floris Books.

Schilthuis, Willy, *Biodynamic Agriculture,* Floris Books.

Soper, John, *Biodynamic Gardening,* Biodynamic Agricultural Association, UK.

Steiner, Rudolf, *Agriculture* (A Course of Eight Lectures), Biodynamic Farming & Gardening Association, USA.

—, *Agriculture: An Introductory Reader,* Steiner Press, UK.

—, *Harmony of the Creative World* (CW 230) Rudolf Steiner Press, UK.

—, *What is Biodynamics? A Way to Heal and Revitalize the Earth,* SteinerBooks, USA

Storl, Wolf, *Culture and Horticulture,* North Atlantic Books.

Thun, Maria, *Gardening for Life,* Hawthorn Press.

—, *The Biodynamic Year,* Temple Lodge.

—, *The Maria Thun Biodynamic Calendar,* Floris Books.

—, *The North American Maria Thun Biodynamic Almanac,* Floris Books.

—, *When Wine Tastes Best: A Biodynamic Calendar for Wine Drinkers,* Floris Books.

von Keyserlink, Adelbert Count, *The Birth of a New Agriculture,* Temple Lodge.

—, *Developing Biodynamic Agriculture,* Temple Lodge.

Waldin, Monty, *Monty Waldin's Best Biodynamic Wines,* Floris Books.

Weiler, Michael, *Bees and Honey, from Flower to Jar,* Floris Books.

Wright, Hilary, *Biodynamic Gardening for Health and Taste,* Floris Books.

Biodynamic Associations

Demeter International
www.demeter.net

Australia:
Bio-Dynamic Research Institute
www.demeter.org.au
Biodynamic Agriculture Australia
www.biodynamics.net.au

Canada: Society for Biodynamic Farming & Gardening
in Ontario
www.biodynamics.on.ca

India: Biodynamic Association of India
www.biodynamics.in

New Zealand: Biodynamic Farming & Gardening Assoc.
www.biodynamic.org.nz

South Africa: Biodynamic Agricultural Association
of Southern Africa
www.bdaasa.org.za

UK: Biodynamic Association
www.biodynamic.org.uk

USA: Biodynamic Association
www.biodynamics.com

Index

NOTES

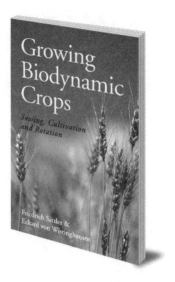

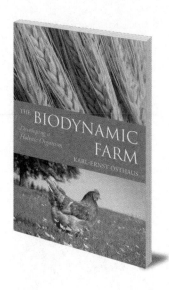

The Biodynamic Farm

Developing a Holistic Organism

Karl-Ernst Osthaus

'*This is more than just a book for the farmer, but rather one for anyone who would like to understand and work with Nature.*'

– SCIENTIFIC & MEDICAL NETWORK REVIEW

Large-scale agriculture tends to view a farm as a means for producing a certain amount of grain, milk or meat. This practical book argues instead for a holistic method of farming: the farm as a living organism. This is the principle of biodynamic farming.

This is an invaluable book for anyone considering setting up a farm, or developing their existing farm with new biodynamic methods.

florisbooks.co.uk

Koepf's Practical Biodynamics

Soil, Compost, Sprays and Food Quality

Herbert H. Koepf

'*Books like this remind us of the parallels between plant and human health, and also all the rhythms of life as a process in time regulated by both the earthly and cosmic environments.*'

– SCIENTIFIC & MEDICAL NETWORK REVIEW

Herbert Koepf was a pioneer of biodynamics in Germany, the USA and in the UK. He was an expert teacher, and drew on his own practical background in farming.

This is an invaluable guide for anyone working with biodynamic methods, offering Koepf's unique insights and wisdom on practical issues.

florisbooks.co.uk

Pfeiffer's Introduction to Biodynamics

Ehrenfried E. Pfeiffer

'*A classic text by one of the earliest biodynamic farmers in North America ... A very useful introduction.*'

– SCIENTIFIC & MEDICAL NETWORK REVIEW

Ehrenfried Pfeiffer was a pioneer of biodynamics in North America. This short but comprehensive book is a collection of three key articles introducing the concepts, principles and practice of the biodynamic method, as well as an overview of its early history.

The book also includes a short biography of Ehrenfried Pfeiffer by Herbert H. Koepf.

florisbooks.co.uk

Weeds and What They Tell Us

Ehrenfried E. Pfeiffer

This wonderful little book covers everything you need to know about the types of plants known as weeds. Ehrenfried Pfeiffer discusses the different varieties of weeds, how they grow and what they can tell us about soil health. The process of combatting weeds is discussed in principle as well as in practice, so that it can be applied to any situation.

florisbooks.co.uk

The Maria Thun Biodynamic Calendar

This useful guide shows the optimum days for sowing, pruning and harvesting various plants and crops, as well as working with bees. It includes Thun's unique insights, which go above and beyond the standard information presented in some other lunar calendars. It is presented in colour with clear symbols and explanations.

The calendar includes a pullout wallchart that can be pinned up in a barn, shed or greenhouse as a handy quick reference.

florisbooks.co.uk

Have you tried our Biodynamic Gardening Calendar app?

A quick, easy way to look up the key sowing and planting information found in the original *Maria Thun Biodynamic Calendar*.

- Filter activities by the time types of the crops you're growing
- Automatically adjusts to your time zone
- Plan ahead by day, week or month
- Available in English, German and Dutch

florisbooks.co.uk

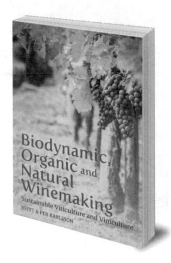

Biodynamic, Organic and Natural Winemaking

Sustainable Viticulture and Viniculture

Britt and Per Karlsson

'The combination of thorough research and personal interviews with growers and winemakers made this material come alive for me.'

– MIKE VESETH, THE WINE ECONOMIST

This comprehensive book by two renowned wine experts explains the rules, the do's and don't's of organic, biodynamic and natural wine production, both outside in the vineyard and in the wine cellar. It sets out clearly what a winemaker is allowed to do, including processes, additives and chemicals, and looks at the potential long-term benefits of going organic or biodynamic.

florisbooks.co.uk

Biodynamic
Wine Growing

Understanding the Vine
and Its Rhythms

Edited by
Jean-Michel Florin

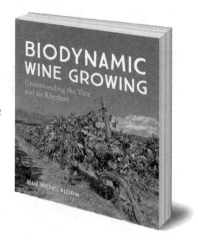

Biodynamic wine expert Jean-Michel Florin has gathered contributions from biodynamic viniculturists to create a beautiful, full-colour book which is both a celebration of sustainable wine growing and an invaluable guide to the future of wine cultivation.

florisbooks.co.uk

Floris
Books

For news on all our **latest books,**
and to receive **exclusive discounts,**
join our mailing list at:

florisbooks.co.uk

Plus subscribers get a FREE book
with every online order!

We will never pass your details to anyone else.